BY APPOINTMENT
TO H.M. QUEEN ELIZABETH THE QUEEN MOTHER
MOTOR CAR MANUFACTURERS
THE DAIMLER COMPANY LIMITED

Daimler

SP250
SPORTS CAR
SPARE PARTS CATALOGUE

ENGINE NUMBER
89001 ONWARDS

CHASSIS NUMBER
100001 ONWARDS

$2\frac{1}{2}$ LITRE V.8 SPORTS CAR

PUBLICATION No. D.5 ISSUED JUNE 1964

THE DAIMLER COMPANY LIMITED, COVENTRY, ENGLAND.

Telephone :
Coventry 27626

Telegraphic Address :
Daimler 'Phone Coventry

INTERPRETATION OF CATALOGUE

Attention is drawn to the lay-out of this catalogue. It will be observed that the heading of each assembly is printed in **BOLD BLOCK LETTERS,** whilst the spares relevant to each assembly are **printed in small type and inset to the right.**

The fact that certain items (printed in small type) are not " inset " indicates that they do not form part of any assembly and must, therefore, be ordered as individual items.

ILLUSTRATIONS

To aid identification of spares, " exploded " illustrations are incorporated in this catalogue.

These illustrations may be used in the following manner:—

(a) If a Crankshaft is required, the index shows that the details for the item are printed on page 4 and that it is illustrated on Plate 2. Turning to page 4, the necessary information for the ordering of the part is given, whilst the column headed " Plate No." indicates that reference to item number 66 on Plate 2 will give a photographic view of the component required.

(b) The details for each illustrated item are printed in the script in numerical Plate number order. An item may therefore be first identified photographically. For example, if a Journal is required for the Propeller Shaft, a picture of the part may be found on Plate No. 18, together with the Plate reference number. It is now an easy matter to find Plate number " 18-8 " in the script and the Part No. and Technical Description are at once apparent.

ORDERING

Customers are advised to make themselves familiar with the method of ordering spare parts in accordance with the explanatory notes which follow.

DESPATCH INSTRUCTIONS

The full address of the destination to which the goods are to be sent, together with that of the customer, should be clearly defined.

Whilst every effort will be made to observe customers' wishes as to the mode of transport, it is the normal custom to utilize the quickest method, whether it be post, passenger train, or road transport, this being dependent upon the size and weight of the package and the destination.

PUBLICATION D.5 PUBLISHED JUNE, 1964

CONTENTS

INSTRUCTIONS FOR ORDERING SPARE PARTS

To avoid delay through inaccurate interpretation, IT IS OF GREAT IMPORTANCE that the information detailed in the following paragraphs be strictly adhered to when placing orders.

CAR TYPE

Always quote the engine and chassis numbers except when Stock Orders are placed, in which case the Year and Model must be indicated.

An order for spares in connection with an Automatic Transmission Unit must be accompanied by the **Transmission Serial Number.** If it is desired to replace the Converter Assembly, however, the order for this component must bear the **Serial number of the Converter** which will be found stamped on the engine side of the Flywheel, adjacent to the Starter Ring Gear.

DETAILS OF PARTS REQUIRED

When ordering, the most complete information should be given to enable us to deal with your requirements expeditiously and to minimise delay in the despatch of the goods. Therefore we strongly recommend that the following procedure be adopted:—

(1) Identify part with technical description in catalogue.

(2) On Order, quote (a) part number, (b) technical description.
NOTE: In no case should the illustration plate number be given.

(3) If doubt exists as to the part number or technical description, a dimensioned sketch will assist in the interpretation of the order, or, alternatively, the old part should accompany the order as a pattern. When patterns are sent, it is **most important** that they be labelled, indicating the name and address of the sender, together with any reference such as order number, etc. Patterns will not be returned unless instructions to that effect accompany the order. In the absence of such instructions, patterns will be scrapped after retention for a period of ONE WEEK.

(4) To order in "sets" is misleading. Please, therefore, **state the specific quantity required.**

SPECIAL ATTENTION

Particular attention is drawn to the fact that when such items as Wings, Doors, Door Glasses, etc., are required, care should be taken to indicate whether they are located on the left-hand or right-hand side of the car. The position is determined by the driver's own left or right hand when seated in the car.

Orders for Carpets, Casings, etc., should clearly state the precise colour required.

DAMAGE BY FIRE OR ACCIDENT

Whilst every precaution is taken, we do not accept liability for damage which may be caused by fire to any motor car or part thereof entrusted to us for any purpose whatsoever. Nor is liability accepted for accidental damage.

INDEX TO PARTS

INDEX TO PARTS

INDEX TO PARTS

INDEX TO PARTS

INDEX TO PARTS

INDEX TO PARTS

Part No.	Description	Plate No.	No. per Unit	Remarks
135890	**CYLINDER BLOCK ASSEMBLY**	1—1	1	Fitted from Engine No. 89001 to 89566
C.21630	**Welch Washer,** at each side of Cylinder Block		6	
C.317	**Welch Washer,** at front of Cylinder Block	1—2	2	
C.21631	**Welch Washer,** at rear of Cylinder Block		2	
260782	**Welch Washer,** at rear end of Camshaft		1	
C.22660	**Plug,** in Oil Gallery	1—3	2	
422583	**Bush,** for Oil Pump Spindle		1	
430381	**Ring Dowel,** locating Cylinder Heads		4	
420304	**Dowel** (Short) at rear of Cylinder Block		1	
428423	**Dowel** (Long) at rear of Cylinder Block		1	
430582	**Liner,** for Camshaft front bearing	1—4	1	
430583	**Liner,** for Camshaft intermediate bearing		4	
430584	**Liner,** for Camshaft rear bearing		1	
307341	**Bearing Cap** (front) for Crankshaft	1—5	1	NOT SUPPLIED SEPARATELY
307340	**Bearing Cap** (intermediate) for Crankshaft	1—6	2	NOT SUPPLIED SEPARATELY
307594	**Bearing Cap** (centre) for Crankshaft	1—7	1	NOT SUPPLIED SEPARATELY
135881	**Bearing Cap** (rear) for Crankshaft	1—8	1	NOT SUPPLIED SEPARATELY
430680	**Dowel,** locating Bearing Caps	1—9	10	
430420	**Bolt,** securing Bearing Caps	1—10	10	
430366	**Lockplate,** securing Bolts	1—11	10	
430386	**Stud,** for Cylinder Heads and Rocker Brackets	1—12	20	
136358	**CYLINDER BLOCK ASSEMBLY**		1	Fitted from Engine No. 89567 to 97540
UCB.137/16R	**Bolt,** securing Rear Bearing Cap		2	Additional Items
FG.106/X	**Washer,** Spring, on Bolts		2	Additional Items
	All other items are as for Cylinder Block Assembly 135890			
C.21034	**CYLINDER BLOCK ASSEMBLY**		1	Fitted to Engine No. 97541 and subs
UCB.137/16R	**Bolt,** securing Rear Bearing Cap		2	Additional items
FG.106/X	**Washer,** Spring, on Bolts		2	Additional items
	All other items are as for Cylinder Block Assembly 135890			
UCS.131/4R	**Plug,** in Cylinder Block water drain holes	1—13	2	
400352	**Copper Washer,** on Plugs	1—14	2	
307476	**TIMING CHAIN COVER**	1—15	1	
C.22611	**Gasket,** between Cover and Cylinder Block	1—16	1	
UCS.125/5R	**Screw,** Set, securing Cover to Cylinder Block		12	
FG.104/X	**Washer,** Spring, on Setscrews		12	
430426	**TIMING CHAIN TENSIONER**	1—17	1	
UCB.125/11R	**Bolt,** securing Tensioner to Cylinder Block	1—18	2	
430421	**Lockplate,** securing Bolts	1—19	1	
307920	**TIMING CHAIN VIBRATION DAMPER**	1—20	1	
UCS.125/6R	**Screw,** Set, securing Damper to Cylinder Block	1—21	2	
400000	**Lockplate,** securing Setscrews	1—22	1	
C.21487	**Screwed Plug,** blanking Chain Oil Feed hole	1—23	1	SUPPLIED WITH CYLINDER BLOCK ASSEMBLY

ENGINE

Part No.	Description	Plate No.	No. per Unit	Remarks
307386	**OIL SUMP ASSEMBLY**	1—25	1	Fitted from Engine No. 89001 to 89566
430378	Dipstick Guide Tube	1—26	1	
308502	**OIL SUMP ASSEMBLY**		1	Fitted to Engine No. 89567 and subs
432069	Dipstick Guide Tube		1	
309039	**Tray Assembly,** in Oil Sump	1—27	1	
UCS.125/4R	**Screw,** Set, securing Tray Assembly to Oil Sump	1—28	5	
422464	**Lockplate,** securing Setscrews	1—29	5	
431285/06	**Drain Plug,** in Oil Sump	1—30	1	
428015	**Copper Washer,** on Drain Plug	1—31	1	
308402	**Oil Seal,** between Bridge Piece and Oil Sump	1—32	1	
308404	**Oil Seal,** between Rear Bearing Cap and Oil Sump	1—33	1	Fitted from Engine No. 89001 to 89566
308470	**Oil Seal,** between Rear Bearing Cap and Oil Sump		1	Fitted to Engine No. 89567 and subs
UCS.131/7R	**Screw,** Set, securing Oil Sump to Cylinder Block		16	
FG.105/X	**Washer,** Spring, on Setscrews		16	
135870	**OIL SUMP BRIDGE PIECE**	1—34	1	
425938	**Bolt,** securing Bridge Piece to Cylinder Block		2	
FG.105/X	**Washer,** Spring, on Bolts		2	
307535	**OIL SUMP DIPSTICK**	1—35	1	
C.22613	**GEARBOX SUPPORT PLATE**	1—36	1	**NOT REQUIRED WHEN AUTOMATIC TRANSMISSION IS FITTED**
C.22614	**GEARBOX SUPPORT PLATE**		1	**REQUIRED WHEN AUTOMATIC TRANSMISSION IS FITTED**
UCS.131/7R	**Screw,** Set, securing Support Plate to Cylinder Block		4	
FG.105/X	**Washer,** Spring, on Setscrews		4	
135893	**CYLINDER HEAD ASSEMBLY**	1—37	2	
C.315	**Welch Washer,** in each end of Cylinder Heads	1—38	4	**NOTE:—On some early Cylinder Heads, Taper Plugs are fitted. When replacement is necessary fit Taper Plug 432099**
431000	**Guide,** for Inlet Valves		8	
431251	**Guide,** for Exhaust Valves		8	
34005/14	**Stud,** for Exhaust Manifold	1—39	16	
430741	**Tube,** for Sparking Plugs	1—40	8	
430742	**Copper Washer,** under Tubes		8	
C.22245	**Gasket,** between Cylinder Heads and Cylinder Block	1—41	2	
UFN.143/L	**Nut,** securing Cylinder Heads and Rocker Brackets to Cylinder Block		20	
C.23586	**Washer,** Plain, under Nuts		20	
UCS.131/4R	**Plug,** blanking Rocker Oil Feed holes in Cylinder Heads	1—42	4	
400352	**Copper Washer,** on Plugs		4	
430444	**Blanking Plate,** sealing Water Apertures in Cylinder Heads	1—43	2	
430429	**Gasket,** between Blanking Plates and Cylinder Heads	1—44	2	
UCS.131/6R	**Screw,** Set, securing Blanking Plates		4	
FG.105/X	**Washer,** Spring, on Setscrews		4	

Part No.	Description	Plate No.	No. per Unit	Remarks
431502	**CYLINDER HEAD TOP COVER**	1—45	2	Fitted from Engine No. 89001 to 89759
307827	**Gasket,** between Top Covers and Cylinder Heads	1—46	2	
431026	**Special Nut,** securing Top Covers to Cylinder Heads	1—47	8	
308620	**CYLINDER HEAD TOP COVER**		2	Fitted to Engine No. 89760 and subs
307827	**Gasket,** between Top Covers and Cylinder Heads		2	
432136	**Special Nut,** securing Top Covers to Cylinder Heads		8	
431308	**Carrier** for Sparking Plug Leads	1—48	8	
431309	**Carrier,** (Centre) for Sparking Plug Lead	1—49	2	
UCS.525/10R	**Screw,** Set, securing Carriers to Top Covers	1—50	2	
UCS.525/7R	**Screw,** Set, securing Carriers to Top Covers	1—51	2	
136014	**OIL FEED PIPE ASSEMBLY FOR VALVE ROCKERS**	1—52	1	
430417	**Banjo Bolt,** securing Oil Feed Pipe to Cylinder Block	1—53	1	
430415	**Banjo Bolt,** securing Oil Feed Pipe to Cylinder Heads	1—54	4	
416490	**Copper Washer,** on Banjo Bolts	1—55	10	
308225	**TAPPET BLOCK ASSEMBLY**		1	
136234	**Tappet Block**	1—56	1	
430388	**Tappet Assembly**		16	
430563	**Ball End** for Tappets		16	
430680	**Dowel,** locating Tappet Block Assembly on Cylinder Block		2	
34005/27	**Stud,** for Dynamo Bracket	1—57	4	
135891	**Gasket,** between Tappet Block and Cylinder Block	1—58	1	
UCS.131/7R	**Screw,** Set, securing Tappet Block to Cylinder Block		10	
UCB.131/10R	**Bolt,** securing Tappet Block to Cylinder Block		2	
FG.105/X	**Washer,** Spring, on Setscrews and Bolts		12	
431288	**OIL FILLER CAP**	1—59	1	
431287	**OIL FILLER TUBE**	1—60	1	
307372	**INDUCTION MANIFOLD ASSEMBLY**	1—61	1	Fitted from Engine No. 89001 to 97553
34005/14	**Stud,** for Carburetters	1—62	8	
C.22412	**INDUCTION MANIFOLD ASSEMBLY**		1	Fitted to Engine No. 97554 and subs.
34005/14	**Stud,** for Carburetters		8	
307876	**Gasket,** between Induction Manifold and Cylinder Heads	1—63	2	
UFN.131/L	**Nut,** securing Carburetters to Induction Manifold Studs		8	
431364	**Plug,** blanking Servo Adaptor hole in Induction Manifold		1	
431363	**Copper Washer,** on Plug		1	
UCS.131/8R	**Screw,** Set, securing Induction Manifold to Cylinder Heads		16	
FG.105/X	**Washer,** Spring, on Setscrews		16	

ENGINE

Part No.	Description	Plate No.	No. per Unit	Remarks
135909	**EXHAUST MANIFOLD**	1—64	2	
307354	**Gasket,** between Exhaust Manifolds and Cylinder Heads	1—65	2	
422804	**Nut,** securing Exhaust Manifolds to Cylinder Head Studs		16	
FW.105/T	**Washer,** Plain, under Nuts		16	
431030	**CRANKSHAFT ASSEMBLY**	2—66	1	Fitted from Engine No. 89001 to 97540
430238	**Dowel,** in Crankshaft Flange	2—67	2	
1216/14	**Bush,** in Crankshaft Flange	2—68	1	
C.22796	**Plug,** in Crankshaft Journals		4	
C.21033	**CRANKSHAFT ASSEMBLY**		1	Fitted to Engine No. 97541 and subs.
430238	**Dowel,** in Crankshaft Flange		2	
1216/14	**Bush,** in Crankshaft Flange		1	
C.22796	**Plug,** in Crankshaft Journals		4	
307359	**Main Bearing** (Front Journal) (Standard)	2—69	2	
307359/1	**Main Bearing** (Front Journal) (·010″ undersize)	2—69	2	
307359/2	**Main Bearing** (Front Journal) (·020″ undersize)	2—69	2	
307359/3	**Main Bearing** (Front Journal) (·030″ undersize)	2—69	2	
307359/4	**Main Bearing** (Front Journal) (·040″ undersize)	2—69	2	
307360	**Main Bearing** (Intermediate Journal, Top) (Standard)	2—70	3	
307360/1	**Main Bearing** (Intermediate Journal, Top) (·010″ undersize)	2—70	3	
307360/2	**Main Bearing** (Intermediate Journal, Top) (·020″ undersize)	2—70	3	
307360/3	**Main Bearing** (Intermediate Journal, Top) (·030″ undersize)	2—70	3	
307360/4	**Main Bearing** (Intermediate Journal, Top) (·040″ undersize)	2—70	3	
308607	**Main Bearing** (Intermediate Journal, Bottom) (Standard)	2—70	3	
308607/1	**Main Bearing** (Intermediate Journal, Bottom) (·010″ undersize)	2—70	3	
308607/2	**Main Bearing** (Intermediate Journal, Bottom) (·020″ undersize)	2—70	3	
308607/3	**Main Bearing** (Intermediate Journal, Bottom) (·030″ undersize)	2—70	3	
308607/4	**Main Bearing** (Intermediate Journal, Bottom) (·040″ undersize)	2—70	3	
307362	**Main Bearing** (Rear Journal) (Standard)	2—71	2	
307362/1	**Main Bearing** (Rear Journal) (·010″ undersize)	2—71	2	
307362/2	**Main Bearing** (Rear Journal) (·020″ undersize)	2—71	2	
307362/3	**Main Bearing** (Rear Journal) (·030″ undersize)	2—71	2	
307362/4	**Main Bearing** (Rear Journal) (·040″ undersize)	2—71	2	
307680/1	**Thrust Washer** (·0920″ thick) for Crankshaft centre Bearing	2—72	2	**ALTERNATIVES**
307680/2	**Thrust Washer** (·0945″ thick) for Crankshaft centre Bearing	2—72	2	
307680/3	**Thrust Washer** (·0970″ thick) for Crankshaft centre Bearing	2—72	2	
431242/1	**Oil Sealing Ring,** at Rear of Crankshaft (·0800″/·0794″) (Standard)	2—73	2	**ALTERNATIVES** Fitted from Engine No. 89001 to 97540
431242/2	**Oil Sealing Ring,** at Rear of Crankshaft (·0794″/·0788″) (Standard)	2—73	2	
431242/3	**Oil Sealing Ring,** at Rear of Crankshaft (·0788″/·0782″) (Standard)	2—73	2	
432751/1	**Oil Sealing Ring,** at Rear of Crankshaft (·0900″/·0894″) (Oversize)	2—73	2	
432751/2	**Oil Sealing Ring,** at Rear of Crankshaft (·0894″/·0888″) (Oversize)	2—73	2	
432751/3	**Oil Sealing Ring,** at Rear of Crankshaft (·0888″/·0882″) (Oversize)	2—73	2	

Part No.	Description	Plate No.	No. per Unit	Remarks
C.22996	**Oil Seal,** at rear of Crankshaft		2	Fitted to Engine No. 97541 and subs.
307407	**FLYWHEEL ASSEMBLY**	2—74	1	**NOT REQUIRED WHEN AUTOMATIC TRANSMISSION IS FITTED**
307406	**Starter Ring Gear**	2—75	1	
420168	**Grubscrew,** securing Ring Gear		6	
430466	**Dowel,** locating Clutch Unit	2—76	2	
430465	**Bolt,** securing Flywheel Assembly to Crankshaft Flange	2—77	6	
430237	**Lockplate,** on Bolts		3	
422510/39	**Key,** locating Crankshaft Chainwheel	2—78	2	
307339	**CHAINWHEEL ON CRANKSHAFT**	2—79	1	
425901	**OIL SHIELD ON CRANKSHAFT**	2—80	1	
431029	**PULLEY ASSEMBLY, ON CRANKSHAFT**	2—81	1	
420304	**Dowel,** locating Pulley	2—82	2	
430912	**Nut,** securing Damper and Pulley to Crankshaft	2—83	1	
425917	**Lockplate,** securing Nut	2—84	1	
307644	**CRANKSHAFT DAMPER**	2—85	1	
UFB.137/12R	**Bolt,** securing Pulley, Damper and Fan	2—86	6	
C.8667/3	**Nut,** Nyloc, on Bolts		6	
136266	**FAN AND EXTENSION ASSEMBLY**	2—87	1	
308642	**Extension Assembly**		1	
136405	**Fan Assembly**		1	
UFB.125/10R	**Bolt,** securing Fan to Extension		4	
430994	**Bush,** Rubber, on Bolts		8	
430995	**Sleeve,** on Bolts		8	
C.8667/1	**Nut,** Nyloc, on Bolts		4	
431206	**Washer,** Plain, under Nuts		4	
307332	**CONNECTING ROD ASSEMBLY**	2—88	8	
307349	**Bearing,** small end	2—89	8	
C.22236	**Bolt,** securing Bearing Caps	2—90	16	
430424	**Nut,** Nyloc, on Bolts		16	
307715	**Bearing** (Big End) (Standard)	2—91	16	
307715/1	**Bearing** (Big End) (·010″ undersize)	2—91	16	
307715/2	**Bearing** (Big End) (·020″ undersize)	2—91	16	
307715/3	**Bearing** (Big End) (·030″ undersize)	2—91	16	
307715/4	**Bearing** (Big End) (·040″ undersize)	2—91	16	

Part No.	Description	Plate No.	No. per Unit	Remarks
308732	**PISTON ASSEMBLY (STANDARD)**	2—92	8	Fitted from Engine No. 89001 to 97371
308864	**PISTON ASSEMBLY (·010″ OVERSIZE)**	2—92	8	
308865	**PISTON ASSEMBLY (·020″ OVERSIZE)**	2—92	8	
308866	**PISTON ASSEMBLY (·030″ OVERSIZE)**	2—92	8	
308867	**PISTON ASSEMBLY (·040″ OVERSIZE)**	2—92	8	
430406	**Pressure Ring** (Upper) Chrome Plated (Standard)	2—93	8	
431517	**Pressure Ring** (Upper) Chrome Plated (·010″ oversize)	2—93	8	
431518	**Pressure Ring** (Upper) Chrome Plated (·020″ oversize)	2—93	8	
431519	**Pressure Ring** (Upper) Chrome Plated (·030″ oversize)	2—93	8	
431520	**Pressure Ring** (Upper) Chrome Plated (·040″ oversize)	2—93	8	
430407	**Pressure Ring** (Lower) Taper (Standard)	2—94	8	
431521	**Pressure Ring** (Lower) Taper (·010″ oversize)	2—94	8	
431522	**Pressure Ring** (Lower) Taper (·020″ oversize)	2—94	8	
431523	**Pressure Ring** (Lower) Taper (·030″ oversize)	2—94	8	
431524	**Pressure Ring** (Lower) Taper (·040″ oversize)	2—94	8	
432209	**Scraper Ring** (Standard)	2—95	8	
432552	**Scraper Ring** (·010″ oversize)	2—95	8	
432553	**Scraper Ring** (·020″ oversize)	2—95	8	
432554	**Scraper Ring** (·030″ oversize)	2—95	8	
432555	**Scraper Ring** (·040″ oversize)	2—95	8	
431449	**Gudgeon Pin,** retaining Pistons on connecting Rods	2—96	8	
430385	**Circlip,** securing Gudgeon Pins in Pistons	2—97	16	
C.21040	**PISTON ASSEMBLY (STANDARD)**		8	Fitted to Engine No. 97372 and subs.
C.21137	**Pressure Ring** (Upper) Chrome Plated		8	
C.21138	**Pressure Ring** (Lower) Taper		8	
C.22111	**Scraper Ring** (Maxiflex '50')		8	
431449	**Gudgeon Pin,** retaining Pistons on Connecting Rods		8	
430385	**Circlip,** securing Gudgeon Pins in Pistons		16	
	THESE PISTONS AND RINGS are obtainable in the following oversizes:— +·010″ +·020″ +·030″ +·040″			
	When ordering Oversize Pistons for C.21040 please quote Part Number and Oversize required (e.g. C.21040/20 for ·020″ oversize)			
C.21282	**CAMSHAFT**	2—98	1	
430365	**Thrust Plate,** at rear end of Camshaft	2—99	1	
UCS.131/5R	**Screw,** Set, securing Thrust Plate to Cylinder Block		2	
C22295	**Lockplate,** securing Setscrews		1	
422510/15	**Key,** locating Chainwheel on Camshaft	2—100	1	
430364	**Thrust Washer,** at front end of Camshaft	2—101	1	
307925	**Chainwheel,** on Camshaft	2—102	1	
430399	**Nut,** securing Chainwheel to Camshaft	2—103	1	
430400	**Lockplate,** securing Nut	2—104	1	

Part No.	Description	Plate No.	No. per Unit	Remarks
307363	**TIMING CHAIN**	2—105	1	
430388	**TAPPET ASSEMBLY**	2—106	16	
430563	**Ball End,** for Tappet	2—107	16	
430754	**INLET PUSH ROD ASSEMBLY**	2—108	8	
430756	**EXHAUST PUSH ROD ASSEMBLY**	2—109	8	
307358	**INLET VALVE**	2—110	8	
307620	**EXHAUST VALVE**	2—111	8	
425910	**Split Cone** (2 Halves)	2—112	16	**USE PART Nos. 432971 AND 432971/1 FOR ALL REPLACEMENTS**
432971	**Split Cone** (Plain Half)		16	
432971/1	**Split Cone** (Padded Half)		16	
430452	**Collar,** at top of Valve Springs	2—113	16	
C.21687	**VALVE SPRING (OUTER)**	2—114	16	
C.21688	**VALVE SPRING (INNER)**	2—115	16	
430980	**Seating Washer,** for Valve Springs	2—116	16	
431000	**VALVE GUIDE (INLET)**	2—117	8	
431251	**VALVE GUIDE (EXHAUST)**	2—118	8	
307368	**ROCKER SHAFT ASSEMBLY (INLET)**	2—119	2	
307378	**ROCKER SHAFT ASSEMBLY (EXHAUST)**	2—120	2	
430425	**ROCKER SHAFT BRACKET ASSEMBLY (INTERMEDIATE)**	2—121	12	
431712	**ROCKER SHAFT BRACKET ASSEMBLY (AT EACH END OF SHAFT)**	2—122	8	
430381	**Dowel,** in Rocker Shaft Brackets	2—123	20	
430551	**Screw,** locating Rocker Shafts in end Brackets	2—124	4	
430552	**Lockplate,** securing Setscrews		4	

Part No.	Description	Plate No.	No. per Unit	Remarks
307352	**VALVE ROCKER (INLET)**	2—125	8	
307353	**VALVE ROCKER (EXHAUST)**	2—126	8	
430411	**Spacing Spring,** between Rockers and Brackets	2—127	16	
430412	**Spacing Washer,** between Rockers, Springs and Brackets	2—128	48	
431011	**Ball End,** on Valve Rockers	2—129	16	
430419	**Nut,** securing Ball End	2—130	16	
307845	**WATER PUMP ASSEMBLY**	3—131	1	
10330	**Pump Body** (18380)		1	
431367	**Bearing,** in Body (18038)	3—132	1	
431368	**Screw,** locating Bearing in Body (18217)	3—133	1	
431369	**Washer,** Fibre, on Locating Screw (18169/2)	3—134	1	
308000	**Impeller,** on Spindle (18391)	3—135	1	
431370	**Seal,** for Bearing (18028)	3—136	1	
431371	**Thrower Ring,** on Spindle (18040)	3—137	1	
307558	**Pulley,** on Spindle	3—138	1	
135916	**WATER PUMP INLET PIPE**	3—139	1	
431172	**Gasket,** between Pump Body and Inlet Pipe	3—140	1	
UCS.131/7R	**Screw,** Set, securing Water Pump to Water Inlet Pipe		5	
FG.105/X	**Washer,** Spring, on Setscrews		5	
430429	**Gasket,** between Water Inlet Pipe Flanges and Cylinder Block	3—141	2	
UCS.131/8R	**Screw,** Set, securing Water Inlet Pipe to Cylinder Block		4	
FG.105/X	**Washer,** Spring, on Setscrews		4	
135938	**WATER OUTLET PIPE, CONNECTING CYLINDER HEADS**	3—142	1	
430429	**Gasket,** between Water Outlet Pipe Flanges and Cylinder Heads	3—143	2	
UCS.131/8R	**Screw,** Set, securing Water Outlet Pipe to Cylinder Heads		4	
FG.105/X	**Washer,** Spring, on Setscrews		4	
304358	**THERMOSTAT UNIT IN WATER OUTLET PIPE (1572115) (OPENS AT 173°F)**	3—144	1	
308080	**WATER TEMPERATURE GAUGE ELEMENT IN WATER OUTLET PIPE (TT. 3800/00)**		1	
307548	**WATER OUTLET PIPE EXTENSION TO RADIATOR TOP HOSE**	3—145	1	
422718	**Gasket,** between Extension and Water Outlet Pipe	3—146	1	
UCS.131/7R	**Screw,** Set, securing Extension to Water Outlet Pipe		2	
FG.105/X	**Washer,** Spring, on Setscrews		2	

Part No.	Description	Plate No.	No. per Unit	Remarks
308074	**WATER PIPE, FROM WATER OUTLET PIPE TO INDUCTION MANIFOLD**	3—147	1	
431495	**Elbow,** Rubber, connecting Water Pipe to Water Outlet Pipe	3—148	1	
431494	**Tee Piece,** Rubber, connecting Water Pipe to Induction Manifold	3—149	1	
422522/04	**Clip,** securing Elbow and Tee Piece		4	
431496	**Plug,** in Tee Piece	3—150	1	NOT REQUIRED WHEN HEATER IS FITTED
422522/03	**Clip,** securing Plug in Tee Piece		1	
308073	**WATER PIPE, FROM WATER PUMP TO TEE PIECE**	3—151	1	
425931	**Hose,** connecting Water Pipe to Water Pump	3—152	1	
422522/04	**Clip,** securing Hose to Water Pipe and Water Pump		2	
431494	**TEE PIECE**	3—153	1	
422522/04	**Clip,** securing Tee Piece to Water Pipes		2	
431496	**Plug,** in Tee Piece	3—154	1	NOT REQUIRED WHEN HEATER IS FITTED
422522/03	**Clip,** securing Plug in Tee Piece		1	
431493	**WATER PIPE, FROM TEE PIECE TO INDUCTION MANIFOLD**	3—155	1	
431495	**Elbow,** Rubber, connecting Water Pipe to Induction Manifold	3—156	1	
422522/04	**Clip,** securing Elbow to Water Pipe and Induction Manifold		2	
C.23528	**OIL PUMP ASSEMBLY**		1	Fitted from Engine No. 89001 to 97664
430360	**Body Assembly**	4—157	1	
430160	**Bush,** in Body	4—158	1	
427569	**Pin,** for Driven Gear, in Body	4—159	1	
430306	**Driving Gear,** for Oil Pump	4—163	1	
430307	**Driven Gear Assembly,** for Oil Pump	4—164	1	
430354	**Cover and Strainer Assembly**	4—165	1	
UCS.125/6R	**Screw,** securing Cover and Strainer	4—166	4	
423044	**Lockplate,** securing Screw	4—167	2	
430362	**Spindle,** for Oil Pump	4—168	1	
422510/11	**Key,** locating Spiral Gear on Spindle	4—169	1	
430363	**Spiral Gear,** on Spindle	4—170	1	
401831	**Thrust Washer,** under Spiral Gear	4—171	1	
C.22756	**Relief Valve,** in Oil Pump Body	4—172	1	
431498	**Spring,** for Relief Valve	4—173	1	
431948	**Cap,** retaining Spring	4—174	1	
431949	**Lockplate,** securing Cap	4—175	1	
C.22490	**OIL PUMP ASSEMBLY**		1	Fitted to Engine No. 97665 and subs.
C.20972	**Driving Gear,** for Oil Pump		1	
C.20971	**Spindle,** for Oil Pump		1	
	All other items are as for Oil Pump Assembly C.23528			

Part No.	Description	Plate No.	No. per Unit	Remarks
430374	**Gasket,** between Oil Pump Body and Rear Bearing Cap	4—160	1	
UCS.131/6R	**Screw,** Set, securing Oil Pump to Rear Bearing Cap	4—161	3	
430356	**Lockplate,** securing Setscrews	4—162	3	
307662	**AIR CLEANER ASSEMBLY**	4—176	2	
UFS.131/6R	**Screw,** Set, securing Air Cleaner to Carburetters		4	
FG.105/X	**Washer,** Spring, on Setscrews		4	
431500	**BREATHER HOSE**	4—177	2	
423176/06	**Clip,** securing Hoses to Air Cleaners and Carburetters		4	
307963	**OIL CLEANER ASSEMBLY (7961049)**		1	
307963/01	**Body Assembly** (complete with Balance Valve) (7961048)	4—178	1	
307963/02	**Gasket,** between Body and Shell (1530764)	4—179	1	
307963/03	**Seal,** at top of Element (1531283)	4—180	1	
307963/04	**Shell Assembly** (7961046)	4—181	1	
307963/03	**Seal,** at bottom of Element (1531283)	4—180	1	
307963/05	**Element** (AC.20)	4—182	1	
10077	**Retainer,** for Element and bottom Seal (1531382)		1	
10078	**Spring,** between Retainer and Shell (1531912)		1	
307963/06	**Bolt,** securing Shell to Body (1530317)	4—183	1	
307963/07	**Washer** (Aluminium) on Bolts (1530710)	4—184	1	
430427	**Gasket,** between Oil Cleaner and Cylinder Block	4—185	1	
UCS.131/8R	**Screw,** Set, securing Oil Cleaner to Cylinder Block		4	
FG.105/X	**Washer,** Spring, on Bolts		4	
431360	**UNION, ON OIL CLEANER FOR OIL PRESSURE PIPE**	4—186	1	
308621	**PIPE ASSEMBLY FOR OIL PRESSURE GAUGE**		1	
431857	**ADAPTOR PLATE FOR OIL CLEANER BODY**		1	Fitted from Engine No. 89001 to 89109 only
431858	**Gasket,** between Adaptor Plate and Oil Cleaner Body		1	Fitted from Engine No. 89001 to 89109 only
136183	**DYNAMO ASSEMBLY (C.40/1—22704)**		1	
136183/01	**Brush** (227541)	5—187	1 set	
136183/02	**Bracket,** commutator end (54211125)	5—188	1	
9704	**Bearing,** for C.E. Bracket (227818)		1	
9017	**Spring,** for Brush Tension (227542)	5—189	1 set	
9705	**Bracket,** driving end (54211095)	5—190	1	
2756	**Bearing,** for D.E. Bracket (189307)		1	
136183/08	**Field Coil** (227543)	5—191	1 set	
136183/09	**Armature** (54214237)	5—192	1	
8586	**Nut,** on Shaft (198800)	5—193	1	

Part No.	Description	Plate No.	No. per Unit	Remarks
308058	**DYNAMO MOUNTING BRACKET ASSEMBLY**	5—194	1	
431972/01	**Shim** (·010" thick) adjusting Bracket		As req'd	
431972/02	**Shim** (·015" thick) adjusting Bracket		As req'd	
UFN.131/L	**Nut,** securing Bracket to Studs on Tappet Block	5—195	4	
431028	**Lockplate,** securing Nuts	5—196	2	
UFB.131/8R	**Bolt,** securing Dynamo to Bracket	5—197	2	
UFN.131/L	**Nut,** on Bolts	5—198	2	
432745	**Washer,** Plain, on Bolts	5—199	2	
FG.105/X	**Washer,** Spring, under Nuts	5—200	2	
UCS.131/7R	**Screw,** Set, securing Dynamo at adjusting Link	5—201	1	
401883	**Serrated Washer,** on Bolt	5—202	1	
FG.105/X	**Washer,** Spring, on Bolt	5—203	1	
431465	**DYNAMO FAN ASSEMBLY**	5—204	1	
307385	**DYNAMO PULLEY**	5—205	1	
307924	**FAN BELT**	5—206	1	
133544	**STARTER MOTOR (M.418G—25533D)**		1	
2804	**Brush** (255659)	5—207	1 set	
2811	**Spring,** for Brush tension (270004)		1 set	
2800	**Armature** (255463)	5—208	1	
2810	**Field Coils** (256578)		1 set	
2801	**Bearing,** for C.E. Bracket (255491)	5—209	1	
6546	**Bracket,** Commutator end (256748)	5—210	1	
6547	**Bracket,** Driving end (256495)		1	
2812	**Bearing,** D.E. Bracket (270038)		1	
133544/07	**Cover Band** (255557)	5—211	1	
133544/09	**Spring,** behind Drive Pinion (255728)	5—212	1	
912672	**Pinion and Sleeve** (255194)	5—213	1	
133544/11	**Main Spring** (270889)	5—214	1	
133544/12	**Nut,** on Shaft (255851)	5—215	1	
UFB.137/18R	**Bolt,** securing Starter Motor		2	
C.8667/3	**Nut,** Nyloc, on Bolts		2	
307670	**DISTRIBUTOR ASSEMBLY (20D8—40676E)**		1	**NOT REQUIRED WHEN AUTOMATIC TRANSMISSION IS FITTED**
9708	**Cap** (54411740)	5—216	1	
6749	**Brush and Spring,** inside Cap (418856)		1	
9709	**Rotor Arm** (425620)	5—217	1	
9711	**Condenser** (54411935)	5—218	2	
9710	**Contact Set** (54411772)	5—219	2	
9713	**Vacuum Unit** (54411675)	5—220	1	
9712	**Springs,** Auto advance (54411934)		1 set	
136104/07	**Cam** (54413267)		1	
136104/08	**Toggle** (407131)		2	
136104/09	**Weight** (54413276)		2	
136104/10	**Shaft and Action Plate** (54413270)		1	
2755	**Bearing** (189272)		1	
6501	**Oil Seal** (418723)		1	

Part No.	Description	Plate No.	No. per Unit	Remarks
	DISTRIBUTOR ASSEMBLY (Cont.)			
9714	**Bush** (423571)		1	
304646	**Ignition Lever** (418190)	5—221	1	
413819	**Washer** (Celeron) under Driving Collar	5—222	1	
431539	**Driving Collar**	5—223	1	
422524/30	**Mills Pin,** retaining Collar		1	
308844	**DISTRIBUTOR ASSEMBLY (20D8—40676E)**		1	**REQUIRED WHEN AUTOMATIC TRANSMISSION IS FITTED**
432513	**Driving Collar**		1	
422524/32	**Mills Pin,** retaining Collar		1	
	All other items are as for Distributor Assembly 307670			
UFS.125/5R	**Screw,** Set, securing Ignition Lever to Mounting Bracket	5—224	2	
FW.104/T	**Washer,** Plain, on Setscrews	5—225	2	
307546	**BRACKET, MOUNTING DISTRIBUTOR**	5—226	1	**NOT REQUIRED WHEN AUTOMATIC TRANSMISSION IS FITTED**
308832	**BRACKET, MOUNTING DISTRIBUTOR**		1	**REQUIRED WHEN AUTOMATIC TRANSMISSION IS FITTED**
UFS.125/7R	**Screw,** Set, securing Bracket to Cylinder Block	5—227	1	
FG.104/X	**Washer,** Spring, on Setscrew	5—228	1	
UFS.131/10R	**Screw,** Set, securing Bracket to Cylinder Block	5—229	1	
FG.105/X	**Washer,** Spring, on Setscrew	5—230	1	
307707	**REV COUNTER PINION**	5—231	1	
431003	**ADAPTOR ASSEMBLY FOR REV COUNTER PINION**	5—232	1	
424789	**Bush,** in Adaptor	5—233	1	
430690	**Screw,** locating Adaptor in Distributor Bracket	5—234	1	
FG.105/X	**Washer,** Spring, on Locating Screw	5—235	1	
431404	**RUBBER RING FOR H.T. CABLES**	5—236	2	
308101	**IGNITION COIL (HA.12—45102/F)**	5—237	1	
UCS.131/4R	**Screw,** Set, securing Coil to Dash		2	
FW.105/T	**Washer,** Plain, on Setscrews		2	
FG.105/X	**Washer,** Spring, on Setscrews		2	
N.8	**SPARKING PLUG (CHAMPION)**	5—238	8	
431262/01	**HIGH TENSION CABLE ASSEMBLY FROM COIL TO DISTRIBUTOR**	5—239	1	

Part No.	Description	Plate No.	No. per Unit	Remarks
	HIGH TENSION SPARKING PLUG LEADS			
432619/17	**Lead to No. 1 R.H. Sparking Plug**	5—240	1	HOME MODELS ONLY 1st CONDITION
432619/12	**Lead to No. 2 R.H. Sparking Plug**	5—240	1	
432619/07	**Lead to No. 3 R.H. Sparking Plug**	5—240	1	
432619/07	**Lead to No. 4 R.H. Sparking Plug**	5—240	1	
432619/12	**Lead to No. 1 L.H. Sparking Plug**	5—240	1	
432619/12	**Lead to No. 2 L.H. Sparking Plug**	5—240	1	
432619/05	**Lead to No. 3 L.H. Sparking Plug**	5—240	1	
432619/05	**Lead to No. 4 L.H. Sparking Plug**	5—240	1	
431252	**HIGH TENSION CABLE**	5—241	As req'd	EXPORT MODELS ONLY 1st CONDITION
425258	**Terminal Washer,** at Lead ends	5—242	10	
425483	**Terminal Nut,** on Distributor	5—243	9	
425257	**Terminal Nut,** on Coil	5—244	1	
432602	**Holder,** at Plug end of Leads	5—245	8	
431704	**Connector,** at Plug end of Leads	5—246	8	
430437	**Ring,** Rubber, for Leads	5—247	8	
	Note:—Use the Assemblies shown below for all H.T. Lead replacements			
	HIGH TENSION SPARKING PLUG LEADS			
432659/10	**Lead to No. 1 R.H. Sparking Plug**		1	ALL MODELS 2nd CONDITION
432659/12	**Lead to No. 2 R.H. Sparking Plug**		1	
432659/06	**Lead to No. 3 R.H. Sparking Plug**		1	
432659/06	**Lead to No. 4 R.H. Sparking Plug**		1	
432659/12	**Lead to No. 1 L.H. Sparking Plug**		1	
432659/06	**Lead to No. 2 L.H. Sparking Plug**		1	
432659/04	**Lead to No. 3 L.H. Sparking Plug**		1	
432659/04	**Lead to No. 4 L.H. Sparking Plug**		1	
432665	**EXTENSION FOR SPARKING PLUG**		8	ALL MODELS 2nd CONDITION
432663	**Contact Clip,** in Extension		8	
432658	**Spring Ring,** for Contact Clip		8	
135996	**CARBURETTERS (HD6)**		1 Pair	
9689	**Body only** (AUC.8243)	6—248	1	L.H. CARBURETTER
135958/01	**Body only** (AUC.8249)	6—249	1	R.H. CARBURETTER
135996/03	**Throttle Spindle** (AUC.4767)	6—250	2	
134640/04	**Throttle Disc** (AUC.3280)	6—251	2	
7046	**Screw,** securing Disc to Spindle (AUC.1358)		4	
135060/30	**Stop,** for Throttle (AUC.2022)	6—252	1	L.H. CARBURETTER
135996/04	**Stop,** for Throttle (AUC.2023)	6—253	1	R.H. CARBURETTER
5485	**Taper Pin,** for Throttle Stop (AUC.2106)		2	
9690	**Lever Assembly,** (AUC.8239)	6—254	2	
5803	**Bolt,** securing Lever to Spindle (AUC.2694)	6—255	2	
5614	**Nut,** on Bolts (AUC.2156)	6—256	2	
9692	**Suction Chamber Assembly** (AUC.8064)	6—257	2	
135996/06	**Piston Spring** (Yellow) (AUC.1167)	6—258	2	
6961	**Damper Assembly** (AUC.8102)	6—259	2	

Part No.	Description	Plate No.	No. per Unit	Remarks
	CARBURETTERS (Cont.)			
5816	**Screw,** securing Suction Chamber to Body (AUC. 2175)	6—260	6	
135996/07	**Jet Needle** (TS)	6—261	2	
9693	**Jet Housing** (AUD.9321)	6—262	1	**L.H. CARBURETTER**
9694	**Jet Housing** (AUD.9320)	6—263	1	**R.H. CARBURETTER**
5780	**Stop Screw,** in Housing (AUC.2521)	6—264	2	
5682	**Spring,** on Stop Screw (AUC.2451)	6—265	2	
6957	**Diaphragm Assembly** (AUC.8155)	6—266	2	
6958	**Jet Bearing** (AUC.2001)	6—267	2	
6959	**Nut,** locking Jet Bearing (AUC.2002)	6—268	2	
6960	**Spring,** under Jet Diaphragm (AUC.2006)	6—269	2	
9697	**Shoe and Rod Assembly** (AUC.8964)	6—270	2	
8760	**Return Spring,** for Shoe and Rod Assembly (AUC. 2020)	6—271	2	
8761	**Plate,** at top of Shoe and Rod Assembly (AUC.2019)	6—273	2	
7879	**Screw,** securing Plate (AUC.4790)	6—274	2	
5780	**Screw,** in Plate, operating Choke/Throttle Control (AUC.2521)	6—275	2	
5682	**Spring,** on Screw (AUC.2451)	6—276	2	
6967	**Valve,** for Slow-running adjustment (AUC.2028)	6—277	2	
6966	**Spring,** for Slow-run Valve (AUC.2027)	6—278	2	
6968	**Gland Washer,** for Slow-run Valve (AUC.2029)	6—279	2	
6969	**Dished Washer** (Brass) at top of Gland Washer (AUC.2030)		2	
7868	**Adaptor,** for Vacuum Suction Pipe (AUC.8190)	6—280	1	**L.H. CARBURETTER**
7091	**Gasket,** under Adaptor (AUC.2014)	6—281	1	**L.H. CARBURETTER**
7961	**Screw,** securing Adaptor (AUC.5156)	6—282	2	**L.H. CARBURETTER**
9695	**Float Chamber** (AUC.2067)	6—283	1	**L.H. CARBURETTER**
9696	**Float Chamber** (AUC.2062)	6—284	1	**R.H. CARBURETTER**
7115	**Bolt,** securing Float Chamber to Carburetters (AUC. 2110)	6—285	8	
1090	**Float** (AUC.1123)	6—286	2	
7118	**Hinged Lever,** operating Float Needle (AUC.1980)	6—287	2	
1153	**Pin,** securing Hinge Lever to Float Chamber Lid (AUC.1152)	6—288	2	
6970	**Needle and Seat,** controlling flow of Petrol to Float Chamber (AUC.8170)	6—289	2	
135996/16	**Lid,** for Float Chamber (AUC.4103)		1	**L.H. CARBURETTER**
9703	**Lid,** for Float Chamber (AUC.4106)	6—290	1	**R.H. CARBURETTER**
7119	**Gasket,** under Float Chamber Lid (AUC.1147)	6—291	2	
1163	**Cap Nut,** securing Lid (AUC.1163)	6—292	2	
7120	**Aluminium Washer,** under Cap Nut (AUC.1557)	6—293	2	
432539	**Ball Pin,** on Mixture Levers		2	
422543/2	**Nut,** securing Ball Pins		2	
FG.104/X	**Washer,** Spring under Nuts		2	
430979	**INSULATOR, BETWEEN CARBURETTERS AND INDUCTION MANIFOLD**	6—295	2	
432974/01	**PETROL PIPE, BETWEEN CARBURETTERS**	6—296	1	Fitted from Engine No. 89001 to 97437
C.20317/01	**PETROL PIPE ASSEMBLY, BETWEEN CARBURETTERS**		1	Fitted to Engine No. 97438 and subs.
C.20295	**Olive,** at connection of Pipe to Carburetters		2	Fitted to Engine No. 97438 and subs.
C.20294	**Union Nut,** securing Pipe to Carburetters		2	Fitted to Engine No. 97438 and subs.

Part No.	Description	Plate No.	No. per Unit	Remarks
136693	**OVERFLOW PIPE, FOR R.H. CARBURETTER**	6—297	1	Fitted from Engine No. 89001 to 97553
136692	**OVERFLOW PIPE, FOR L.H. CARBURETTER**		1	Fitted from Engine No. 89001 to 97553
432975	**CORAGRIP PIPE CONNECTING OVERFLOW PIPES TO CARBURETTERS**		2	Fitted from Engine No. 89001 to 97553
431354	**Clip,** securing Overflow Pipes		4	Fitted from Engine No. 89001 to 97553
C.22384	**OVERFLOW PIPE, FOR R.H. CARBURETTER**		1	Fitted to Engine No. 97544 and subs.
C.22385	**OVERFLOW PIPE, FOR L.H. CARBURETTER**		1	Fitted to Engine No. 97544 and subs.
C.22386	**CORAGRIP PIPE CONNECTING OVERFLOW PIPE TO CARBURETTERS**		2	Fitted to Engine No. 97544 and subs.
431354	**Clip,** securing Overflow Pipes		4	Fitted to Engine No. 97544 and subs.
307991	**SUCTION PIPE ASSEMBLY**	6—298	1	
422519/01	**Nipple,** on Pipe		1	
422520/01	**Union Nut,** on Pipe		1	
C.14212	**Connector,** Rubber, connecting Suction Pipe to L.H. Carburettor	6—299	1	

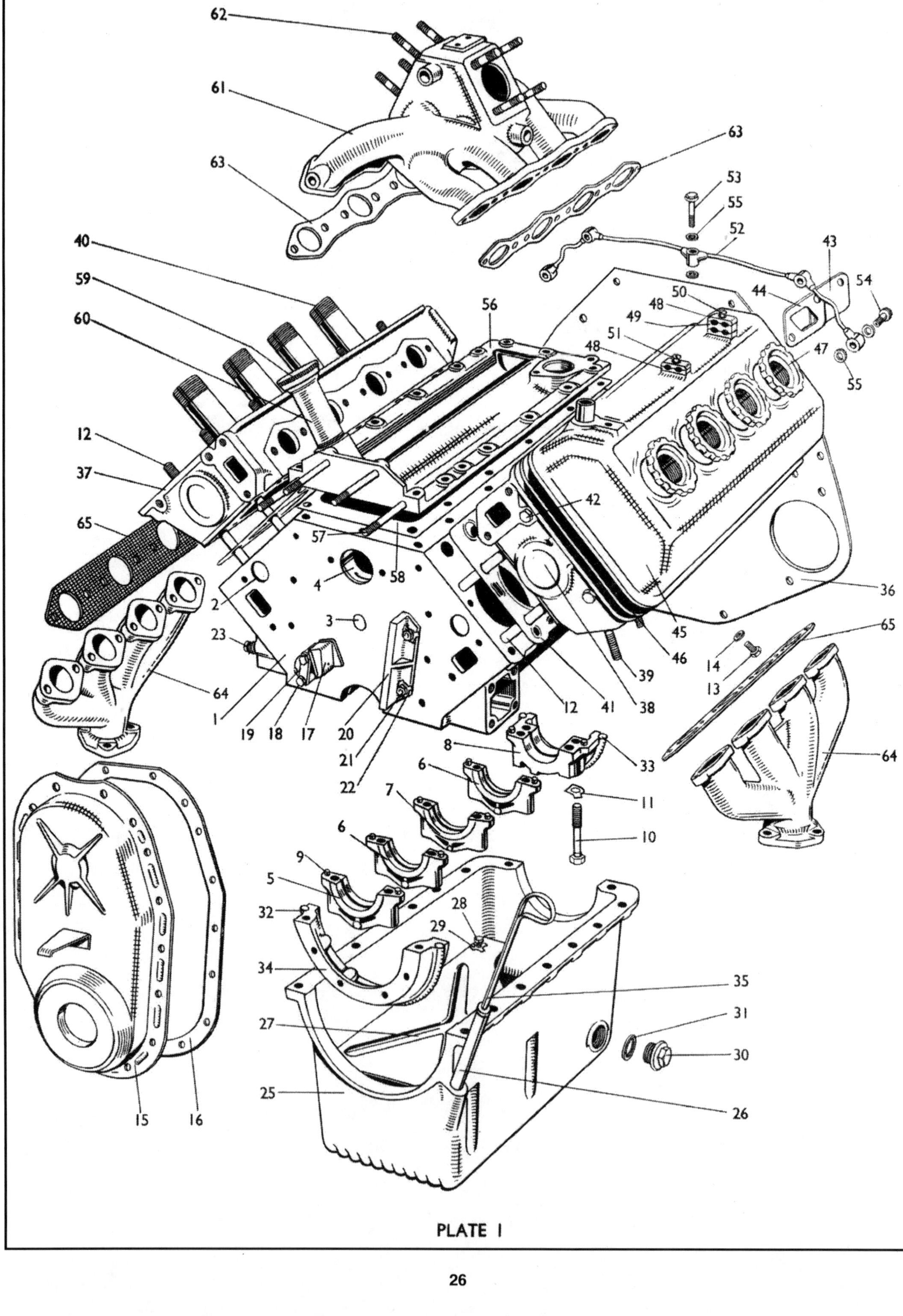
62
61
63
63
53
55
52
43
54
44
40
59
60
56
50
48
49
51
48
47
55
12
37
42
65
57
58
4
2
36
45
3
23
65
14
13
46
39
64
1
19
18
17
20
21
22
12
41
38
8
33
64
6
11
7
10
6
9
5
32
28
29
34
35
31
30
27
25
26
15
16

PLATE 1

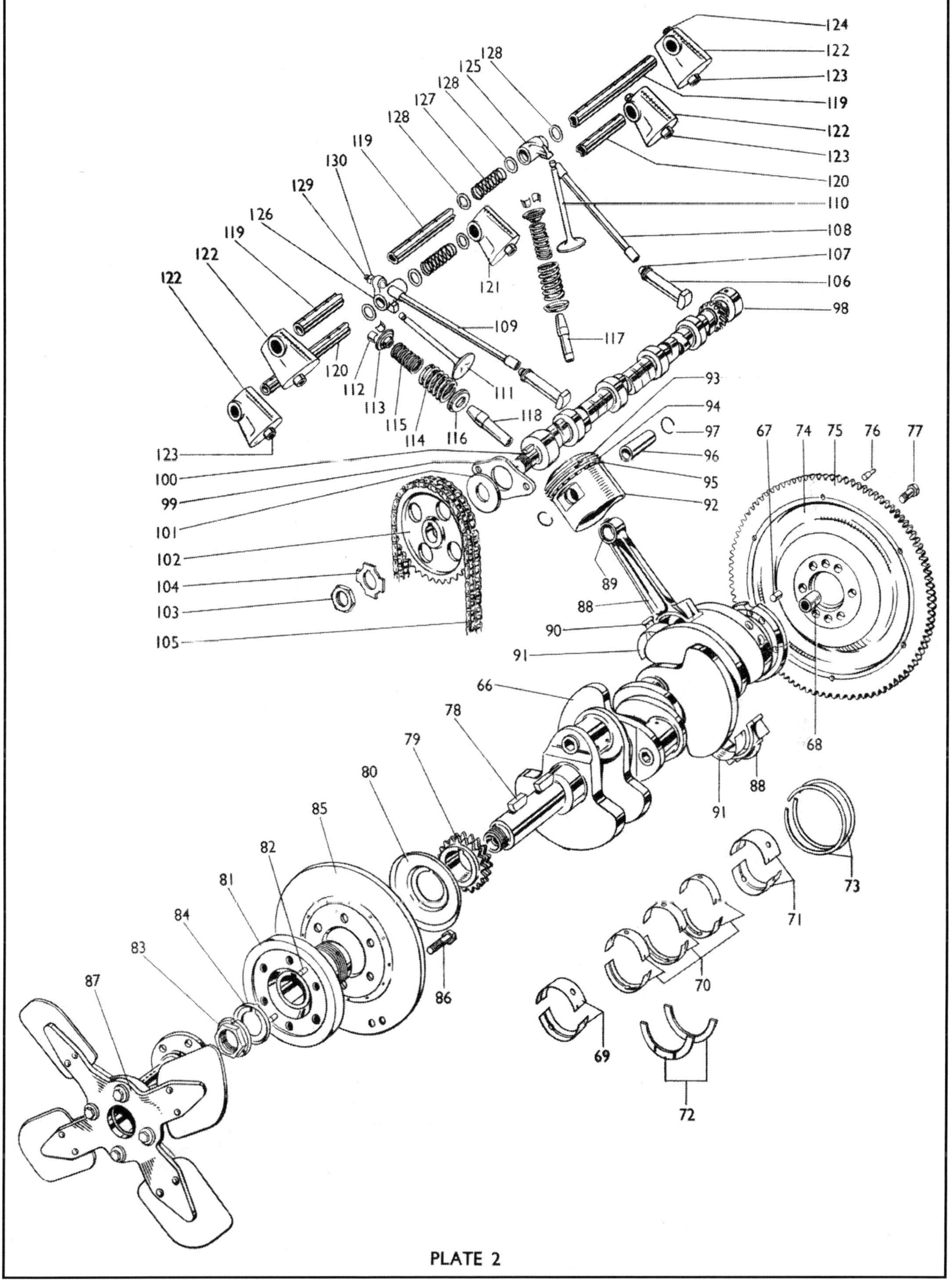
124
122
123
119
122
123
120
110
108
107
106
98
128
125
128
127
128
119
130
129
126
119
122
122
121
109
117
120
112
113
111
115
114
116
118
93
94
97
96
95
92
123
100
99
101
102
104
103
105
67
74
75
76
77
89
88
90
91
66
78
79
80
85
82
81
84
83
87
86
68
88
91
73
71
70
69
72

PLATE 2

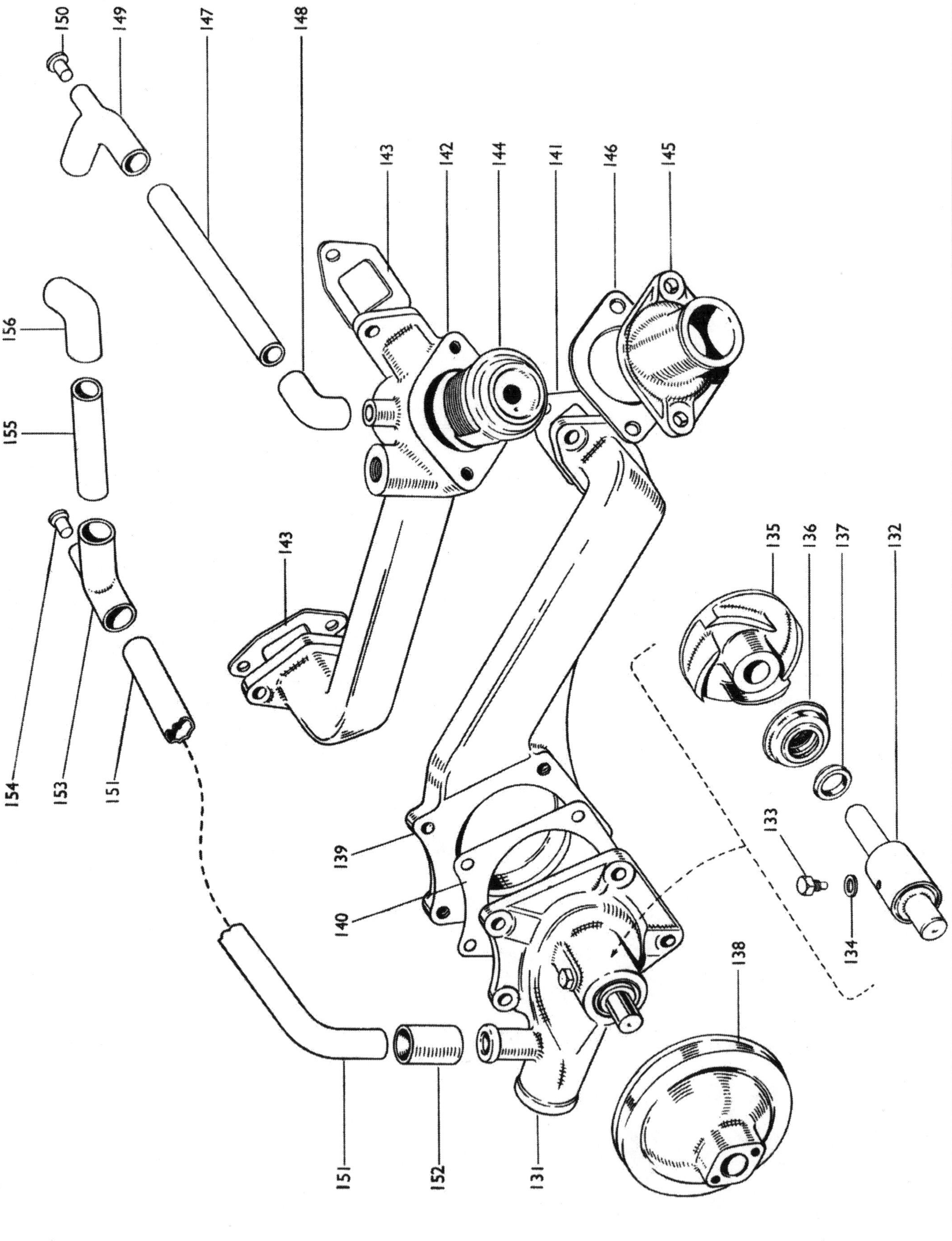
150
149
147
148
143
142
144
141
146
145
156
155
143
135
136
137
132
154
153
151
133
139
140
134
138
151
152
131

PLATE 3

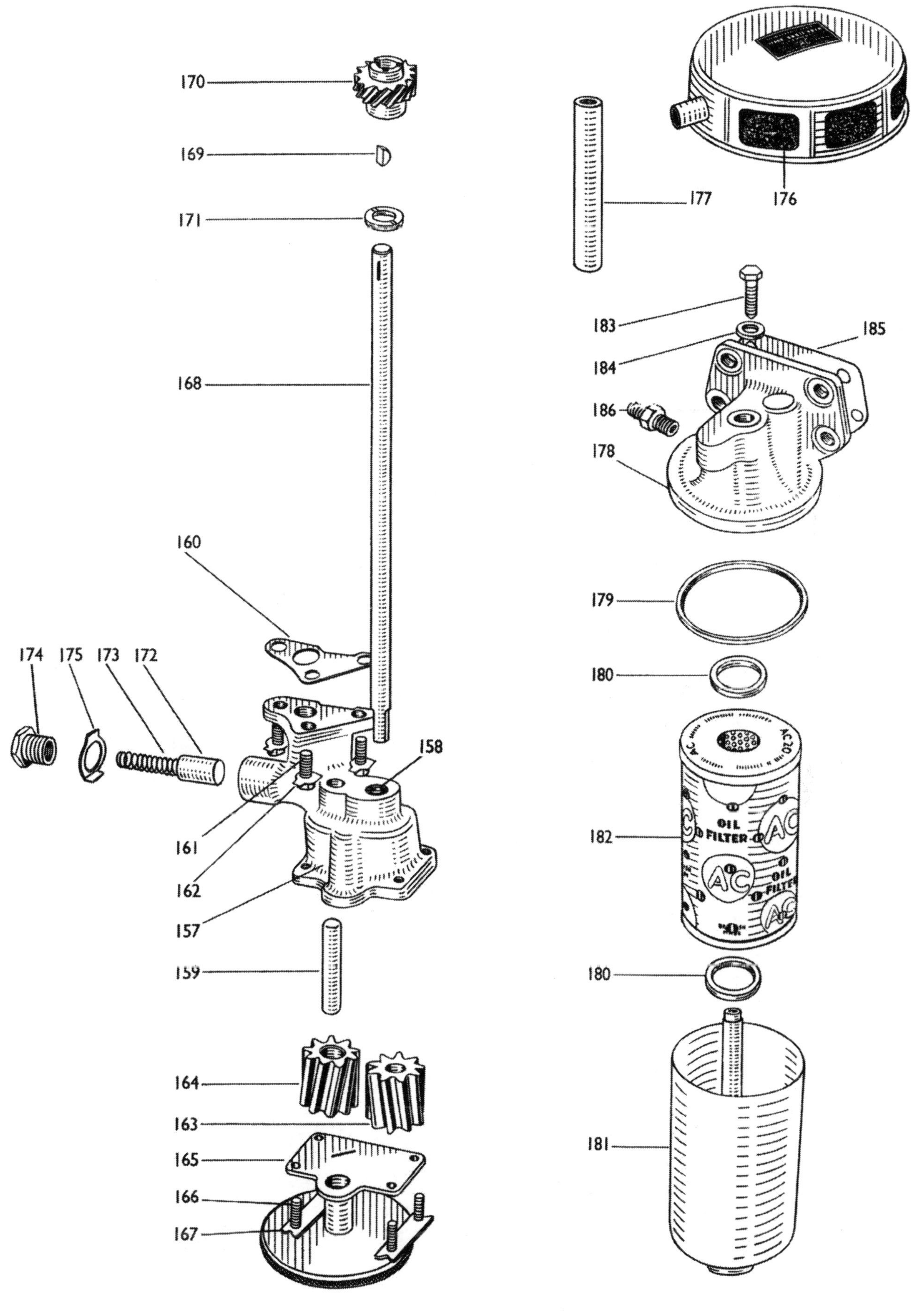
170
169
171
168
160
174
175
173
172
158
161
162
157
159
164
163
165
166
167
177
176
183
184
185
186
178
179
180
182
OIL FILTER
AC
180
181

PLATE 4

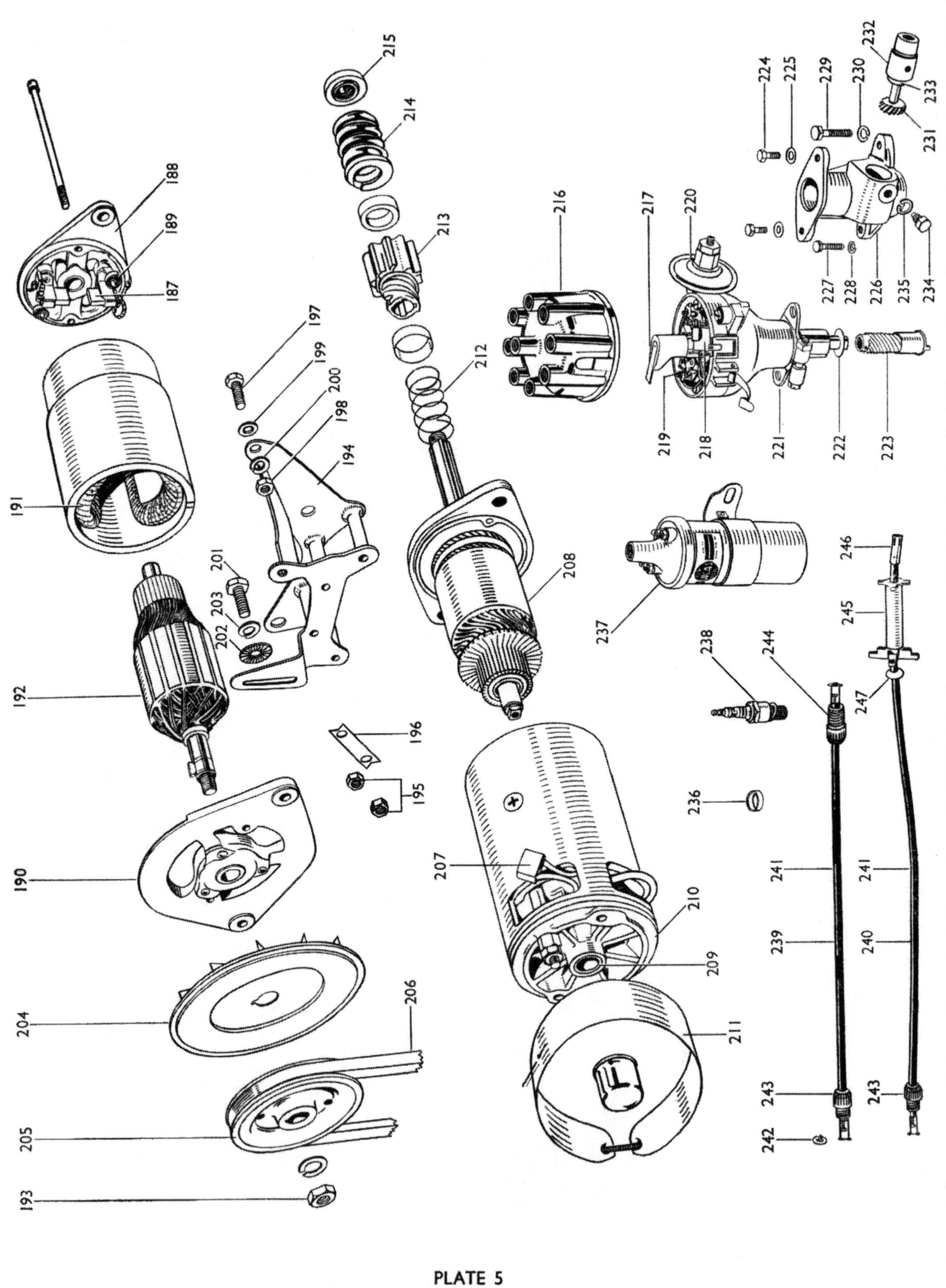
187
188
189
190
191
192
193
194
195
196
197
198
199
200
201
202
203
204
205
206
207
208
209
210
211
212
213
214
215
216
217
218
219
220
221
222
223
224
225
226
227
228
229
230
231
232
233
234
235
236
237
238
239
240
241
241
242
243
243
244
245
246
247

PLATE 5

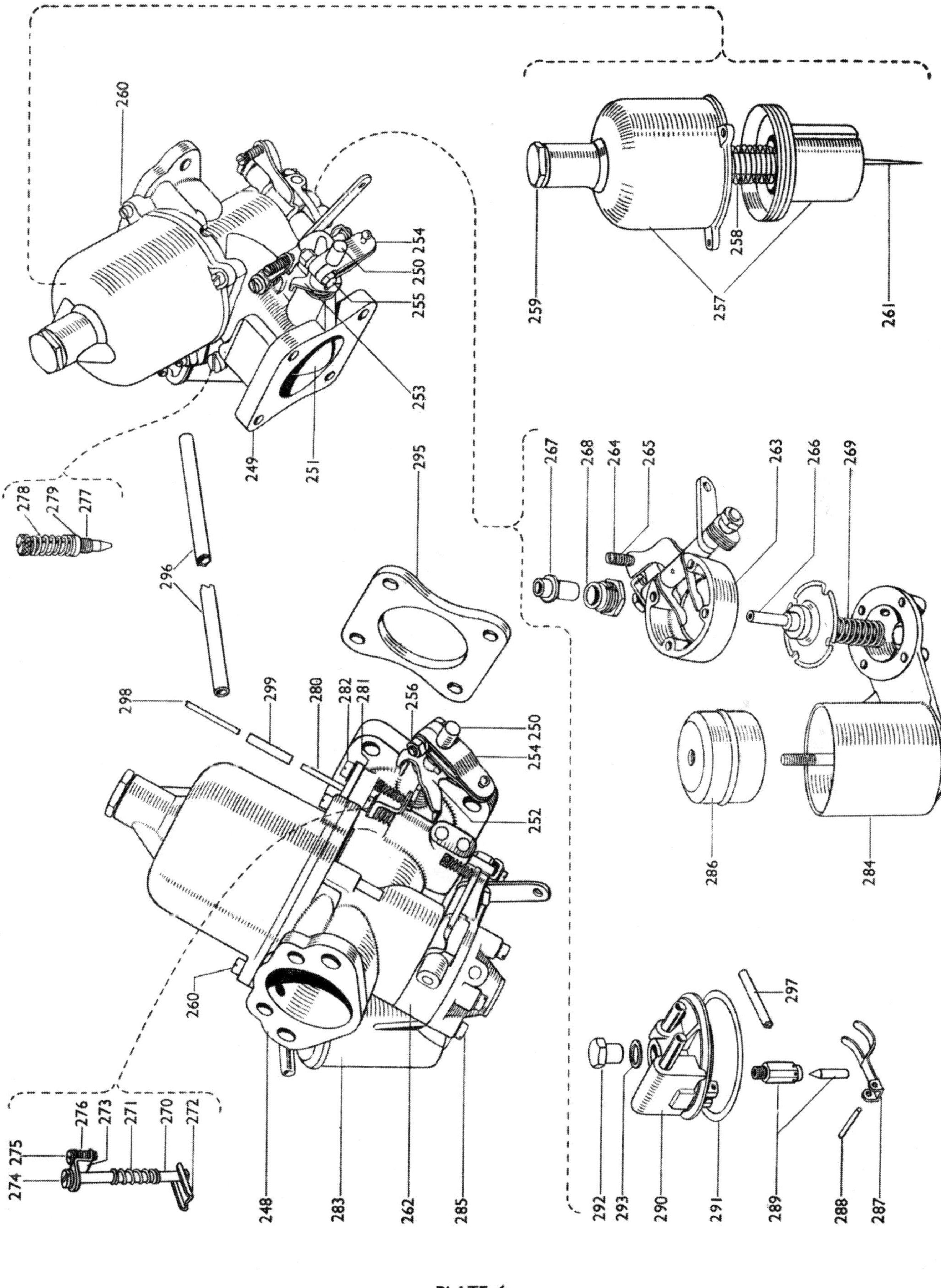

260
254
250
255
253
251
249
259
258
257
261
278
279
277
296
295
267
268
264
265
263
266
269
298
299
280
282
281
256
250
254
252
286
284
260
297
274 275
276
273
271
270
272
248
283
262
285
292
293
290
291
289
288
287

PLATE 6

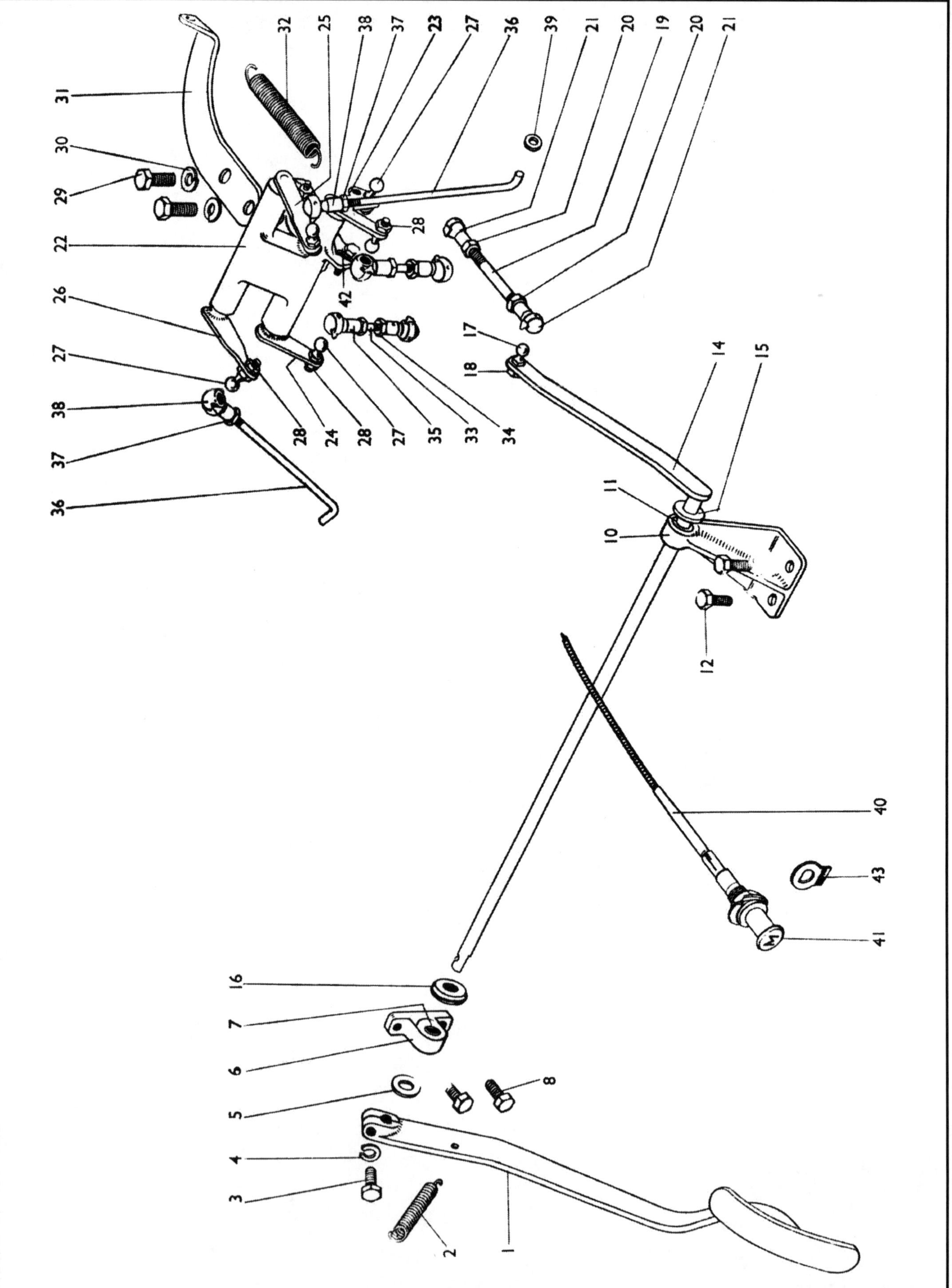
32
25
38
37
23
27
36
39
21
20
19
20
21
31
30
29
22
26
27
38
37
36
28
42
28
24
28
27
35
33
34
17
18
14
15
11
10
12
40
43
41
16
7
6
5
8
4
3
2
1

PLATE 7

ACCELERATOR CONTROLS
(STANDARD TRANSMISSION)

Part No.	Description	Plate No.	No. per Unit	Remarks
307930	**ACCELERATOR PEDAL ASSEMBLY**		1	**R.H. DRIVE CARS ONLY**
308025	**ACCELERATOR PEDAL ASSEMBLY**	7—1	1	**L.H. DRIVE CARS ONLY**
431507	**Return Spring,** for Pedal Assembly	7—2	1	
UFS.125/6R	**Screw,** Set, clamping Pedal Assembly to Cross Shaft	7—3	1	
FG.104/X	**Washer,** Spring, on Setscrew	7—4	1	
FW.104/T	**Washer,** Plain, between Pedal and Bracket	7—5	1	
432037	**STOP FOR ACCELERATOR PEDAL**		1	**R.H. DRIVE CARS ONLY**
UFS.125/5R	**Screw,** Set, securing Pedal Stop		2	
UFN.125/L	**Nut,** on Setscrews		2	
FW.104/T	**Washer,** Plain, on Setscrews		2	
FG.104/X	**Washer,** Spring, under Nuts		2	
432046	**STOP FOR ACCELERATOR PEDAL**		1	**L.H. DRIVE CARS ONLY**
90018/04	**Bolt,** securing Pedal Stop		2	
430684	**BRACKET ASSEMBLY (OUTER) FOR ACCELERATOR CROSS SHAFT**	7—6	1	
0608/08	**Bush,** in Bracket	7—7	1	
90018/06	**Bolt,** securing Bracket to Dash	7—8	2	
431419	**BRACKET ASSEMBLY (INNER) FOR ACCELERATOR CROSS SHAFT**	7—10	1	
0608/08	**Bush,** in Bracket	7—11	1	
90018/06	**Bolt,** securing Bracket to Dash	7—12	2	
431286	**CROSS SHAFT ASSEMBLY FOR ACCELERATOR PEDAL**		1	**R.H. DRIVE CARS ONLY**
431413	**CROSS SHAFT ASSEMBLY FOR ACCELERATOR PEDAL**	7—14	1	**L.H. DRIVE CARS ONLY**
FW.106/T	**Washer,** Plain, between Lever on Cross Shaft and Inner Bracket	7—15	As req'd	
432277	**Grommet,** in Dash for Cross Shaft	7—16	1	
424813	**Ball Pin,** on Lever	7—17	1	
UFN.119/L	**Nut,** securing Ball Pin	7—18	1	
AG.102/X	**Washer,** Spring, under Nut		1	

ACCELERATOR CONTROLS
(STANDARD TRANSMISSION)

Part No.	Description	Plate No.	No. per Unit	Remarks
431418	**THROTTLE CONTROL ROD, FROM CROSS SHAFT LEVER TO CONTROL BRACKET**	7—19	1	
UFN.125/L	**Nut,** on Control Rod securing Ball Socket	7—20	2	
420963/3	**Ball Socket,** on Control Rod	7—21	2	
307761	**THROTTLE AND MIXTURE CONTROL BRACKET ASSEMBLY**	7—22	1	
431065	**R.H. Throttle Lever and Spindle**	7—23	1	
431068	**L.H. Throttle Lever**	7—24	1	
431061	**R.H. Mixture Lever and Spindle**	7—25	1	
431063	**L.H. Mixture Lever**	7—26	1	
422524/08	**Mills Pin,** for L.H. Levers		2	
424813	**Ball Pin,** on Throttle and Mixture Levers	7—27	5	
UFN.119/L	**Nut,** securing Ball Pins	7—28	5	
AG.102/X	**Washer,** Spring, under Nuts		5	
UCS.131/8R	**Screw,** Set, securing Control Bracket to Induction Manifold	7—29	2	
FG.105/X	**Washer,** Spring, on Setscrews	7—30	2	
431417	**Anchorage,** for Throttle Return Spring	7—31	1	
428770	**THROTTLE RETURN SPRING**	7—32	1	
431097	**THROTTLE CONTROL ROD FROM CONTROL BRACKET TO CARBURETTERS**	7—33	2	
UFN.119/L	**Nut,** on Control Rods, securing Ball Sockets	7—34	4	
424822/3	**Ball Socket,** on Control Rods	7—35	4	
431074	**MIXTURE CONTROL ROD FROM CONTROL BRACKET TO CARBURETTERS**	7—36	2	
UFN.119/L	**Nut,** on Control Rods securing Ball Sockets	7—37	2	
424822/3	**Ball Socket,** on Control Rods	7—38	2	
427461	**Washer,** Plain, behind Split Pins	7—39	2	
L.102/5U	**Split Pin,** securing Control Rods to Levers on Carburetters		2	
307759/05	**MIXTURE CONTROL CABLE ASSEMBLY**	7—40	1	
431607	**Knob,** inscribed 'M', on Cable	7—41	1	
432279	**Grommet,** in Dash, for Control Cable		1	
405315	**Clip,** securing Control Cable	7—42	1	
426933/35	**Screw,** securing Clip		1	
426935/4	**Nut,** on Screw		1	
AG.102/X	**Washer,** Spring, under Nut		1	
431529	**Designation Plate,** "MIXTURE"	7—43	1	

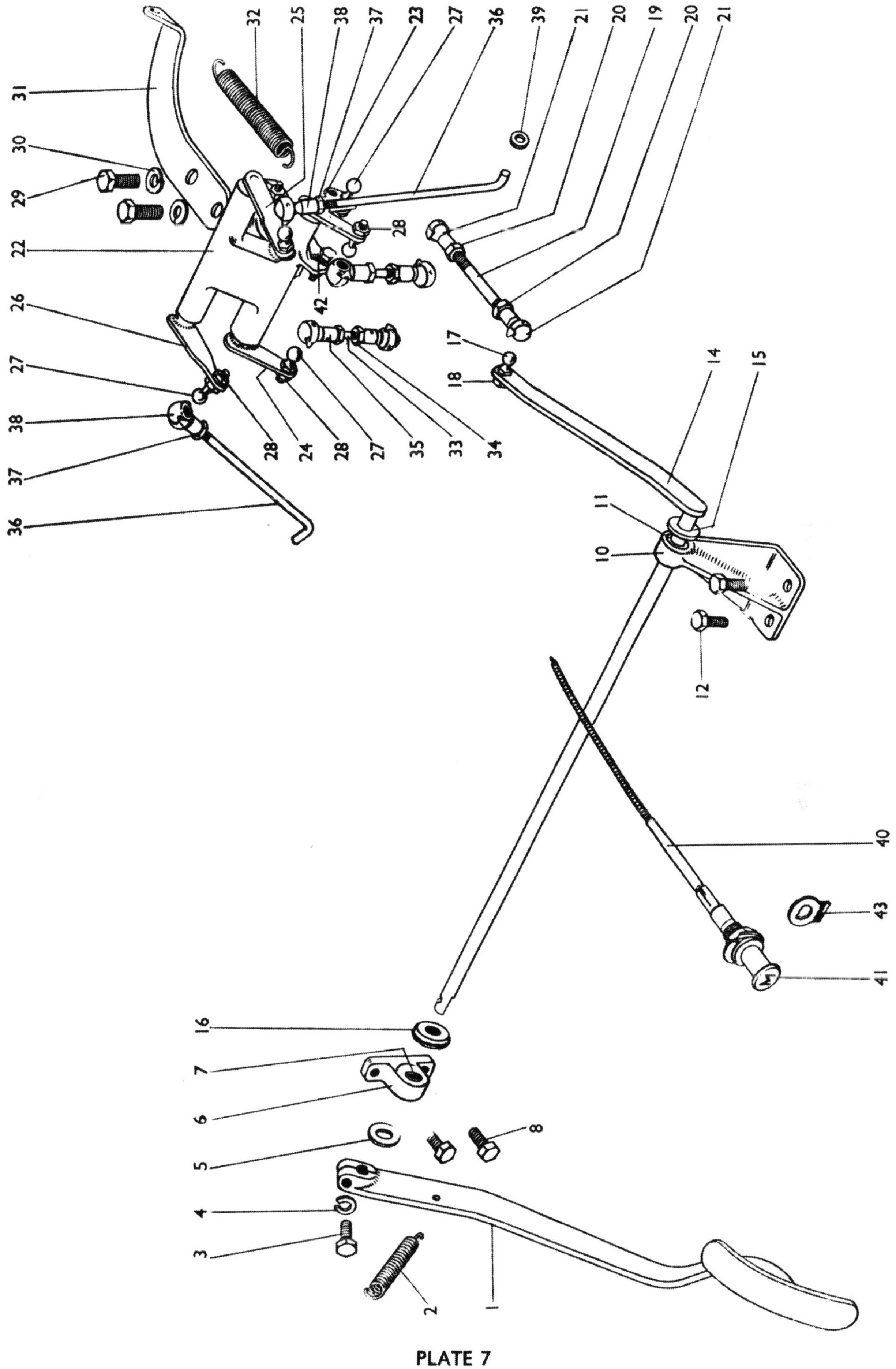

31
30
29
22
26
27
38
37
36
32
25
38
37
23
27
36
39
21
20
19
20
21
28
42
28
24
28
27
35
33
34
17
18
14
15
11
10
12
40
43
41
16
7
6
5
8
4
3
2
1

PLATE 7

ACCELERATOR CONTROLS
(AUTOMATIC TRANSMISSION)

Part No.	Description	Plate No.	No. per Unit	Remarks
308809	**ACCELERATOR PEDAL ASSEMBLY**	8—1	1	**R.H. DRIVE CARS ONLY**
308807	**ACCELERATOR PEDAL ASSEMBLY**		1	**L.H. DRIVE CARS ONLY**
431507	**Return Spring,** for Pedal Assembly	8—2	1	
UFS.125/6R	**Screw,** Set, clamping Pedal Assembly to Cross Shaft		1	
FG.104/X	**Washer,** Spring, on Setscrews		1	
FG.104/T	**Washer,** Plain, between Pedal and Bracket		1	
432932	**STOP, FOR ACCELERATOR PEDAL**	8—3	1	**R.H. DRIVE CARS ONLY**
UFS.125/5R	**Screw,** Set, securing Pedal Stop		2	
UFN.125/L	**Nut,** on Setscrews		2	
FW.104/T	**Washer,** Plain, on Setscrews		2	
FG.104/X	**Washer,** Spring, under Nuts		2	
432933	**STOP, FOR ACCELERATOR PEDAL**		1	**L.H. DRIVE CARS ONLY**
90018/04	**Bolt,** securing Pedal Stop		2	
432489	**BRACKET (OUTER) FOR ACCELERATOR CROSS SHAFT**	8—4	1	
90018/06	**Bolt,** securing Bracket to Dash		2	
432532	**BRACKET (INNER) FOR ACCELERATOR CROSS SHAFT**	8—5	1	
90018/06	**Bolt,** securing Bracket to Dash		2	
432931	**STIFFENER ASSEMBLY FOR CROSS SHAFT INNER BRACKET**	8—6	1	
UFS.125/5R	**Screw,** Set, securing Stiffener		2	
FG.104/X	**Washer,** Spring, on Setscrews		2	
432469	**CROSS SHAFT ASSEMBLY FOR ACCEL-ERATOR PEDAL**	8—7	1	**R.H. DRIVE CARS ONLY**
432474	**CROSS SHAFT ASSEMBLY FOR ACCEL-ERATOR PEDAL**		1	**L.H. DRIVE CARS ONLY**
FW.107/T	**Washer,** Plain, between Lever on Cross Shaft and Inner Bracket		1	
432928	**Spring,** for Accelerator Cross Shaft	8—8	1	
432534	**Grommet,** in Dash, for Cross Shaft	8—9	1	
424813	**Ball Pin,** on Lever on Cross Shaft	8—10	1	
UFN.119/L	**Nut,** securing Ball Pin		1	
AG.102/X	**Washer,** Spring, under Nut		1	

Part No.	Description	Plate No.	No. per Unit	Remarks
431418	**THROTTLE CONTROL ROD, FROM CROSS SHAFT LEVER TO CONTROL BRACKET**	8—11	1	
UFN.125/L	**Nut,** on Control Rod, securing Ball Socket		2	
420936/3	**Ball Socket,** on Control Rod	8—12	2	
308726	**THROTTLE AND MIXTURE CONTROL BRACKET ASSEMBLY**	8—13	1	
431795	**Spring,** Throttle Relay		1	
424813	**Ball Pin,** on Throttle and Mixture Levers	8—14	5	
UFN.119/L	**Nut,** securing Ball Pins		5	
AG.102/X	**Washer,** Spring, under Nuts		5	
UCS.131/8R	**Screw,** Set, securing Control Bracket to Induction Manifold		2	
FG.105/X	**Washer,** Spring, on Setscrews		2	
431417	**Anchorage,** for Throttle Return Spring	8—15	1	
428770	**THROTTLE RETURN SPRING**	8—16	1	
431097	**THROTTLE CONTROL ROD FROM CONTROL BRACKET TO CARBURETTERS**	8—17	2	
UFN.119/L	**Nut,** on Control Rods, securing Ball Sockets		4	
424822/3	**Ball Socket,** on Control Rods	8—18	4	
431862	**MIXTURE CONTROL ROD, FROM CONTROL BRACKET TO CARBURETTER**	8—19	2	
UFN.119/L	**Nut,** on Control Rods, securing Ball Sockets		2	
424822/3	**Ball Socket,** on Control Rods	8—20	2	
427461	**Washer,** Plain, behind Split Pins		2	
L.102/5U	**Split Pin,** securing Control Rods to Levers on Carburetters		2	
422542/15	**KICK-DOWN CONTROL ROD, FROM THROTTLE AND MIXTURE CONTROL BRACKET TO LEVER ON TAPPET BLOCK**	8—21	1	
420963/3	**Ball Socket,** on Control Rod	8—22	2	
UFN.125/L	**Nut,** securing Ball Sockets		2	
424813/2	**Ball Pin,** on Control Bracket and Kick-down Levers	8—23	2	
UFN.119/L	**Nut,** securing Ball Pins to Levers		2	
AG.102/X	**Washer,** Spring, under Nuts		2	

ACCELERATOR CONTROLS
(AUTOMATIC TRANSMISSION)

Part No.	Description	Plate No.	No. per Unit	Remarks
432256	**KICK-DOWN LEVER ASSEMBLY, ON REAR OF TAPPET BLOCK**	8—24	1	
432264	**Fulcrum Pin,** securing Kick-down Lever to Support Bracket	8—25	1	
C.8667/1	**Nut,** Nyloc, securing Lever to Fulcrum Pin		1	
FW.104/T	**Washer,** Plain, under Nut		1	
308706	**SUPPORT BRACKET FOR KICK-DOWN LEVER**	8—26	1	
UCS.131/5R	**Screw,** Set, securing Bracket to Tappet Block		1	
FG.105/X	**Washer,** Spring, on Setscrew		1	
422542/20	**CONTROL ROD, FROM LEVER ON TAPPET BLOCK TO LEVER ON BELL-HOUSING**	8—27	1	
420963/3	**Ball Socket,** on Control Rod	8—28	2	
UFN.125/L	**Nut,** securing Ball Sockets to Rod		2	
424813/2	**Ball Pin,** on Levers	8—29	2	
UFN.119/L	**Nut,** securing Ball Pins		2	
AG.102/X	**Washer,** Spring, under Nuts		2	
432252	**KICK-DOWN LEVER ASSEMBLY, ON BELL HOUSING**	8—30	1	
432253	**Collar,** on Spindle of Kick-down Lever Assembly	8—31	1	
422523/09	**Mills Pin,** securing Collar to Spindle		1	
422542/45	**CONTROL ROD, FROM LEVER ON BELL HOUSING TO SELECTOR LEVER ON TRANSMISSION UNIT**	8—32	1	
420963/3	**Ball Socket,** on Control Rod	8—33	2	
UFN.125/L	**Nut,** securing Ball Sockets to Rod		2	
424813/2	**Ball Pin,** on Levers	8—34	2	
UFN.119/L	**Nut,** securing Ball Pins to Levers		2	
AG.102/X	**Washer,** Spring, under Nuts		2	
307759/13	**MIXTURE CONTROL CABLE ASSEMBLY**	8—35	1	
431607	**Knob,** on Cable		1	
432279	**Grommet,** in Dash for Control Cable	8—36	1	
405315	**Clip,** securing Control Cable		1	
426933/35	**Screw,** securing Clip		1	
426935/4	**Nut,** on Screw		1	
AG.102/X	**Washer,** Spring, under Nut		1	
431529	**Designation Plate,** "MIXTURE"	8—37	1	

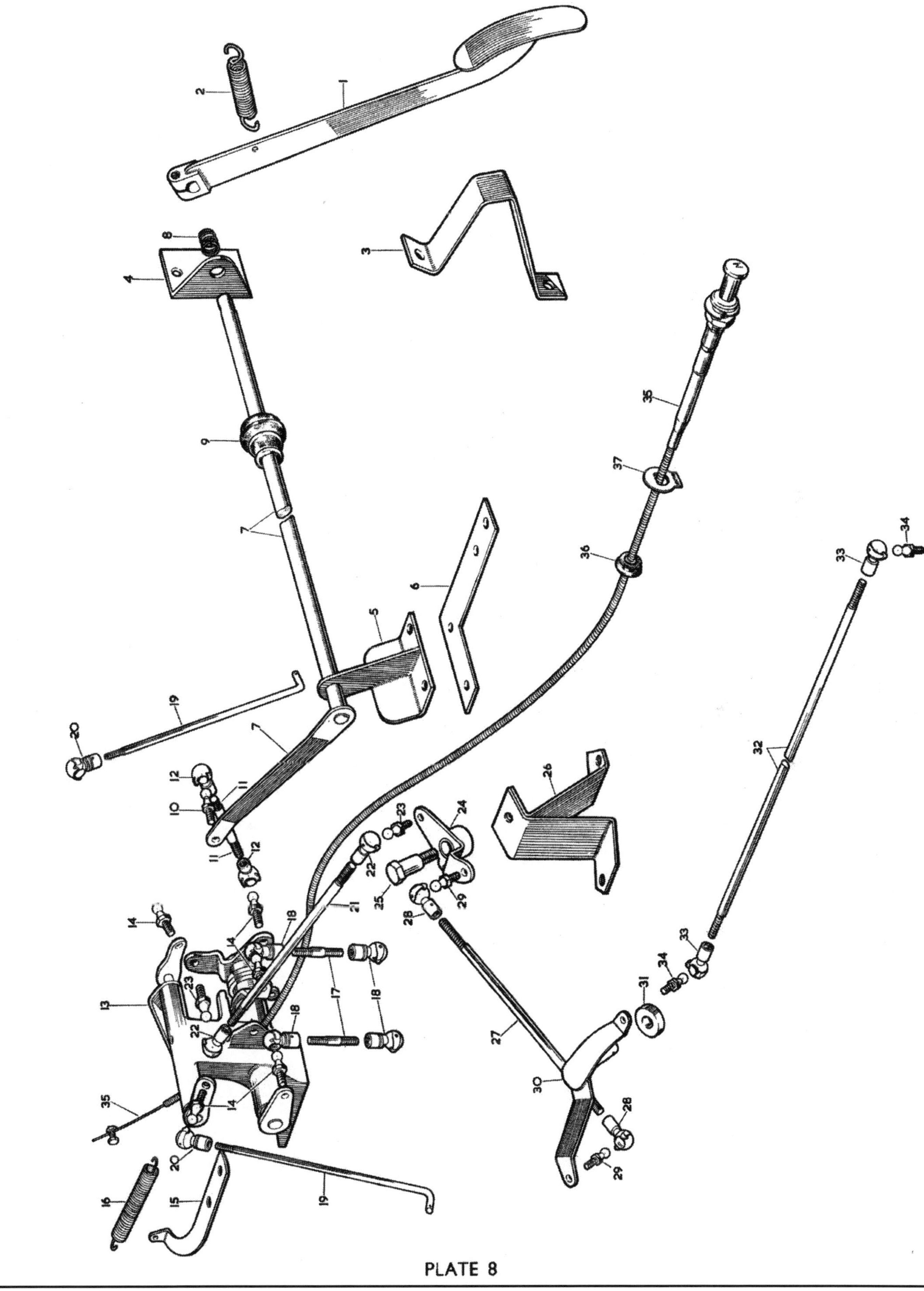
1
2
3
4
8
9
7
6
5
36
37
35
19
20
7
12
11
10
11
12
26
24
23
22
25
29
28
21
18
14
14
13
23
17
18
22
18
17
18
14
35
20
16
15
19
27
31
34
33
30
28
29
32
33
34

PLATE 8

CLUTCH UNIT AND CONTROLS

STANDARD TRANSMISSION ONLY

Part No.	Description	Plate No.	No. per Unit	Remarks
431333	**CLUTCH PEDAL ASSEMBLY**	9—1	1	**R.H. DRIVE CARS ONLY**
1012/11	**Bush,** in Pedal	9—2	2	
431398	**CLUTCH PEDAL ASSEMBLY**		1	**L.H. DRIVE CARS ONLY**
1012/11	**Bush,** in Pedal	9—2	2	
427000	**Rubber Pad,** on Clutch Pedal	9—3	1	
413046	**Return Spring,** for Clutch Pedal	9—4	1	
UFS.125/6R	**Screw,** Set, for adjustment of Clutch Pedal		1	
UFN.225/L	**Locknut,** on Setscrew		1	
136118	**BRACKET MOUNTING CLUTCH AND BRAKE PEDALS**		1	
90018/06	**Screw,** Set, securing Bracket to Dash		2	
UFS.125/6R	**Screw,** Set, securing Bracket to Dash		4	
51/637052	**Washer,** Plain, on Setscrews		4	
C.8667/2	**Nut,** Nyloc, on Setscrews		4	
430635	**FULCRUM SHAFT FOR CLUTCH AND BRAKE PEDALS**		1	
L.104/6U	**Split Pin,** retaining Shaft in Pedal Mounting Bracket		2	
FW.110/T	**Washer,** Plain, behind Split Pin		2	
430637	**Spacer,** between Clutch and Brake Pedals		1	
308048	**CLUTCH MASTER CYLINDER ASSEMBLY (64067612)**	9—5	1	
308048/01	**Filler Cap,** on Reservoir (378434)	9—6	1	
305167/03	**Gasket,** for Filler Cap (362920)	9—7	1	
308048/04	**Spring,** on Valve (64673425)	9—8	1	
308048/02	**Push Rod** (350527W)	9—9	1	
306086/03	**Retaining Washer** (378230)	9—10	1	
308260/07	**CLUTCH MASTER CYLINDER REPAIR KIT (SP.2095)**	9—11	1	
	Comprising:—			
	Valve Seal			
	Plunger Seal			
	Dust Cover			
	Circlip			
	Valve Spring Washer			
	Filler Cap Gasket			
	Tube of Grease			
307973	**Distance Piece,** between Clutch and Brake Master Cylinders and Dash		1	
UFB.131/24R	**Bolt,** securing Clutch Master Cylinder to Dash	9—12	2	
C.8667/2	**Nut,** on Bolts	9—13	2	

Part No.	Description	Plate No.	No. per Unit	Remarks
430934	**JAW, CONNECTING CLUTCH MASTER CYLINDER TO CLUTCH PEDAL**	9—14	1	
UFN.231/L	**Locknut,** securing Jaw to Master Cylinder	9—15	1	
422531/16	**Clevis Pin,** securing Jaw to Pedal	9—16	1	
L.103/5U	**Split Pin,** retaining Clevis Pin		1	
FW.105/T	**Washer,** Plain, behind Split Pin	9—17	1	
136172	**PIPE ASSEMBLY, FROM MASTER CYLINDER TO HOSE**		1	**R.H. DRIVE CARS ONLY**
136173	**PIPE ASSEMBLY, FROM MASTER CYLINDER TO HOSE**	9—18	1	**L.H. DRIVE CARS ONLY**
305172	**FLEXIBLE HOSE FROM PIPE ASSEMBLY TO SLAVE CYLINDER (3700619W)**	9—19	1	
426751	**Nut,** securing Hose to Chassis (10233)	9—20	1	
425163	**Washer,** Shakeproof under Nut (74330)	9—21	1	
431441	**Gasket,** on Hose (378711)	9—22	1	
308049	**CLUTCH SLAVE CYLINDER ASSEMBLY (3010320W)**	9—23	1	
308049/01	**Spring,** behind Piston (378802)	9—24	1	
135347/11	**Bleeder Screw** (377686W)	9—25	1	
134617/05	**Dust Cap,** on Bleeder Screw (363400)	9—26	1	
308049/02	**SLAVE CYLINDER REPAIR KIT**	9—27	1	
	Comprising:—			
	Seal			
	Dust Cover			
	Bleeder Screw Dust Cap			
	Tube of Grease			
307414	**BRACKET, SUPPORTING SLAVE CYLINDER**		1	
UCS.131/6R	**Screw,** Set, securing Bracket to Gearbox		1	
FG.105/X	**Washer,** Spring, on Setscrews		1	
UFS.131/7R	**Screw,** Set, securing Slave Cylinder to Bracket		2	
UFN.131/L	**Nut,** on Setscrews		2	
FG.105/X	**Washer,** Spring, under Nuts		2	
431319	**PUSH ROD FOR SLAVE CYLINDER**	9—28	1	
431318	**Jaw,** connecting Push Rod to Cross Shaft Lever	9—29	1	
UFN.231/L	**Locknut,** securing Jaw to Push Rod	9—30	1	
J.105/13S	**Clevis Pin,** securing Jaw to Cross Shaft Lever	9—31	1	
L.103/5U	**Split Pin,** retaining Clevis Pin		1	
FW.105/T	**Washer,** Plain, behind Split Pin	9—32	1	

CLUTCH UNIT AND CONTROLS

STANDARD TRANSMISSION ONLY

Part No.	Description	Plate No.	No. per Unit	Remarks
307977	**CLUTCH CROSS SHAFT ASSEMBLY**	9—33	1	
430470	**Return Spring,** on Cross Shaft	9—34	1	
UCS.131/18R	**Screw,** Set, locating Cross Shaft	9—35	1	
FG.105/X	**Washer,** Spring on Setscrew	9—36	1	
431950	**Grease Nipple,** on Cross Shaft	9—37	1	
307976	**CLUTCH OPERATING FORK**	9—38	1	
430459	**Screw,** locating Operating Fork on Cross Shaft	9—39	1	
136139/01	**CLUTCH COVER ASSEMBLY (45692/84)**		1	
2556	**Cover only** (45481)	9—40	1	
2582	**Spring,** Anti-rattle (47688)	9—41	3	
136139/04	**Pressure Plate** (48572)	9—42	1	
7834	**Thrust Spring** (White) (40123)	9—43	9	
2529	**Release Lever** (42662)	9—45	3	
2472	**Plate,** for Release Lever (40968)	9—46	1	
2395	**Retainer,** for Release Lever (40015)	9—47	3	
2527	**Strut,** for Release Lever (42606)	9—48	3	
2523	**Pin,** for Release Lever (42604)	9—49	3	
2588	**Eye Bolt,** for Release Lever (48243)	9—50	3	
5086	**Nut,** on Eyebolt (48132)	9—51	3	
136139/14	**CLUTCH DRIVEN PLATE ASSEMBLY (47626/110)**	9—52	1	
2590	**CLUTCH RELEASE BEARING ASSEMBLY (46890)**	9—53	1	
4163	**Retainer,** for Release Bearing (41628)	9—54	1	
UFS.131/6R	**Screw,** Set, securing Clutch to Flywheel		6	
FG.105/X	**Washer,** Spring, on Setscrews		6	

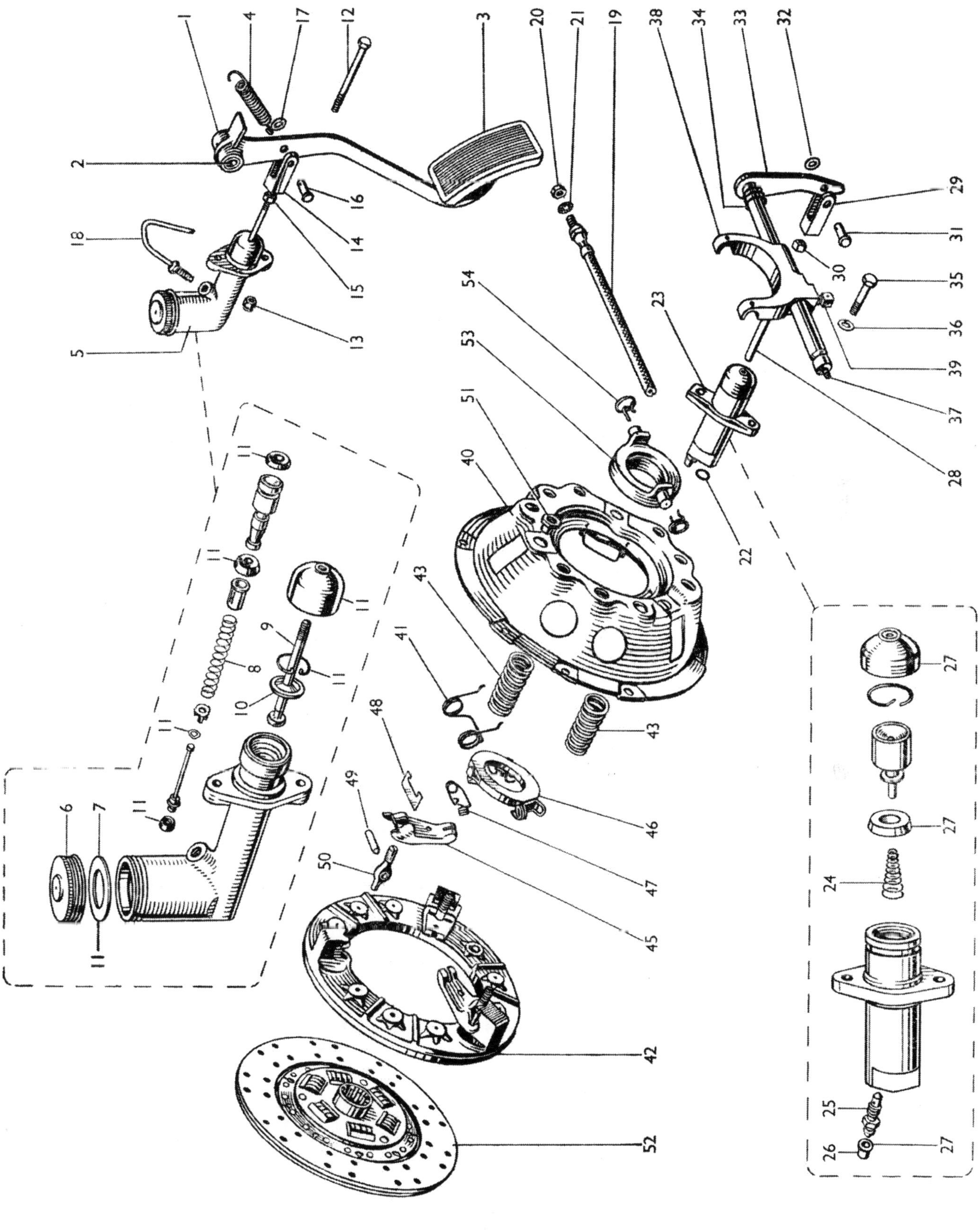

1
4
17
12
3
20
21
19
38
34
33
32
2
16
14
15
18
13
5
54
53
51
40
23
29
31
30
35
36
39
37
28
22
11
11
11
11
9
8
10
11
11
11
6
7
11
43
41
43
48
49
50
46
47
45
42
52
27
27
24
25
26
27

PLATE 9

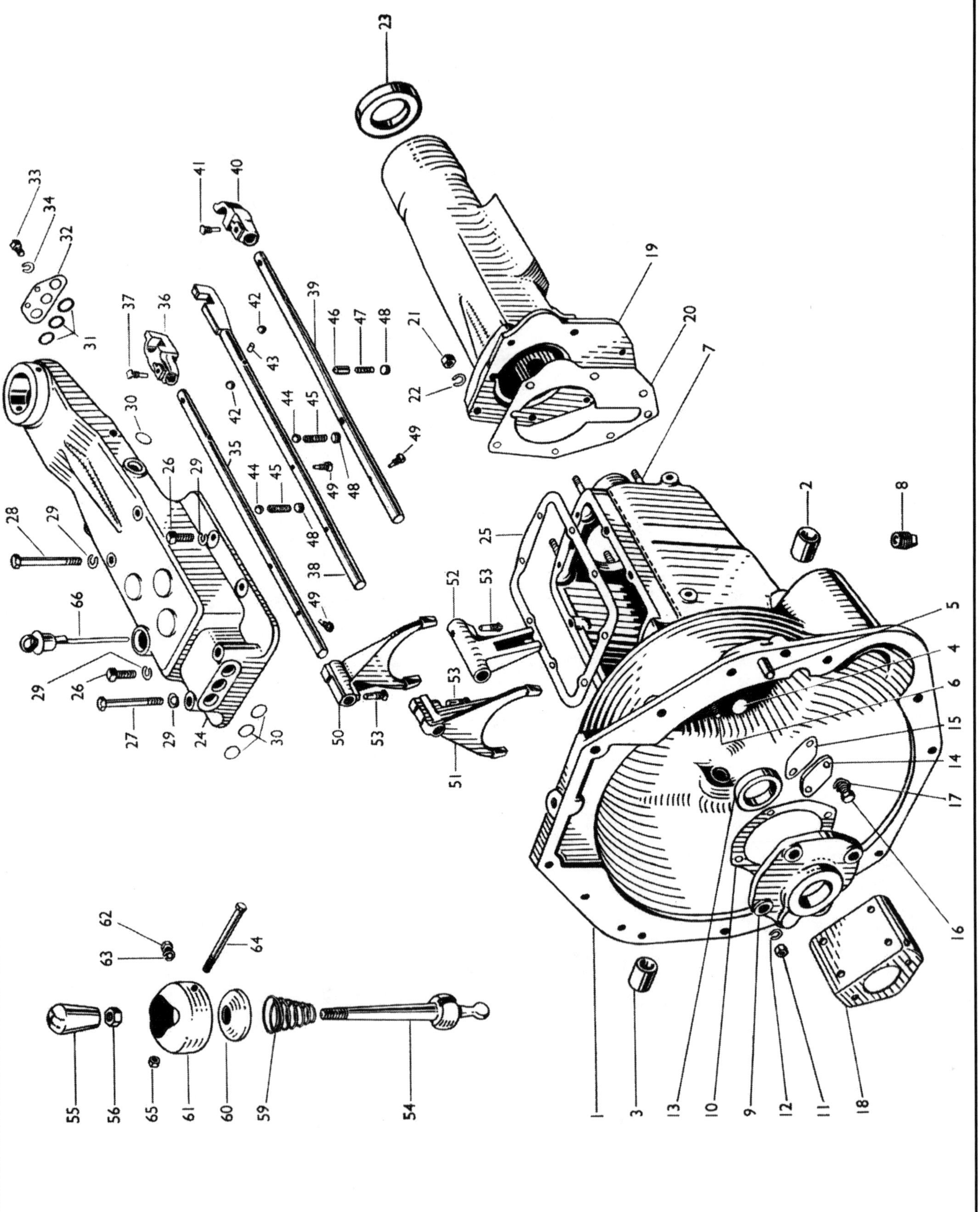

1
2
3
4
5
6
7
8
9
10
11
12
13
14
15
16
17
18
19
20
21
22
23
24
25
26
27
28
29
30
31
32
33
34
35
36
37
38
39
40
41
42
43
44
45
46
47
48
49
50
51
52
53
54
55
56
59
60
61
62
63
64
65
66

PLATE 10

GEARBOX (SYNCHROMESH)
STANDARD TRANSMISSION ONLY

Part No.	Description	Plate No.	No. per Unit	Remarks
DL.36833	**GEARBOX ASSEMBLY COMPLETE**		1	
136101	**GEARBOX AND CLUTCH CASING ASSEMBLY**	10—1	1	
430463	**L.H. Bush,** for Clutch Cross Shaft	10—2	1	
1214/16	**R.H. Bush,** for Clutch Cross Shaft	10—3	1	
266365	**Welch Washer,** in reverse gear operating rod access hole	10—4	1	
420304	**Dowel,** locating Flywheel Housing	10—5	1	
34005/12	**Stud,** for Front End Cover	10—6	4	
34005/14	**Stud,** for Rear Casing	10—7	7	
430510	**Drain Plug,** in Casing	10—8	1	
307978	**FRONT END COVER**	10—9	1	
430514	**Gasket,** between Front End Cover and Casing	10—10	1	
UFN.131/L	**Nut,** securing Front End Cover to Casing	10—11	4	
FG.105/X	**Washer,** Spring, under Nuts	10—12	4	
423200/02	**Oil Seal,** for Primary Shaft	10—13	1	
430495	**End Cover,** for Countershaft	10—14	1	
430496	**Gasket,** between End Cover and Casing	10—15	1	
UCS.131/5R	**Screw,** Set, securing End Cover	10—16	2	
FG.105/X	**Washer,** Spring, on Setscrews	10—17	2	
307414	**BRACKET, MOUNTING SLAVE CYLINDER**	10—18	1	
UCS.131/6R	**Screw,** Set, securing Bracket to Gearbox		4	
FG.105/X	**Washer,** Spring, on Setscrews		4	
135915	**REAR CASING FOR GEARBOX**	10—19	1	
307440	**Gasket,** between Rear Casing and Gearbox Casing	10—20	1	
UFN.131/L	**Nut,** securing Rear Casing	10—21	7	
FG.105/X	**Washer,** Spring, under Nuts	10—22	7	
405011	**Thrust Piece,** for Speedometer Pinion		1	
431696	**Oil Seal,** for Mainshaft	10—23	1	
136112	**TOP COVER FOR GEARBOX**	10—24	1	
307426	**Gasket,** between Top Cover and Casing	10—25	1	
UFS.131/8R	**Screw,** Set, securing Top Cover	10—26	4	
UCB.131/22R	**Bolt,** securing Top Cover	10—27	2	
UCB.131/24R	**Bolt,** securing Top Cover	10—28	2	
FG.105/X	**Washer,** Spring, on Setscrews and Bolts	10—29	8	
266365	**Welch Washer,** in Top Cover	10—30	5	
422503/06	**Plug,** in Top Cover for Interlock access		1	
425280/07	**Sealing Ring,** for Selector Rod	10—31	1	
430476	**Sealing Ring,** for End Cover	10—32	1	
UCS.125/4R	**Screw,** Set, securing Top Cover to End Cover	10—33	2	
FG.104/X	**Washer,** Spring, on Setscrews	10—34	2	

GEARBOX (SYNCHROMESH)
STANDARD TRANSMISSION ONLY

Part No.	Description	Plate No.	No. per Unit	Remarks
307897	**SELECTOR ROD, 1st/2nd GEARS**	10—35	1	
307416	**Selector,** on 1st/2nd Gear Selector Rod	10—36	1	
430475	**Screw,** Taper, securing Selector to Rod	10—37	1	
307900	**SELECTOR ROD ASSEMBLY, 3rd/4th GEARS**	10—38	1	
307898	**SELECTOR ROD, REVERSE GEAR**	10—39	1	
307417	**Selector,** on Reverse Selector Rod	10—40	1	
430475	**Screw,** Taper, securing Selector to Rod	10—41	1	
425292/11	**Ball** ($\frac{3}{8}''$) for Selector Rod Interlock	10—42	2	
430479	**Plug,** for Selector Rod Interlock	10—43	1	
425292/11	**Ball,** for forward Selector Rod Indent	10—44	2	
432598	**Spring,** for Indent Balls	10—45	2	
430505	**Thimble,** for Reverse Selector Rod	10—46	1	
430508	**Spring,** for Thimble	10—47	1	
430506	**Screw,** retaining Spring	10—48	3	
430474	**Stop Screw,** for Selector Rods	10—49	3	
307433	**SELECTOR FORK, 1st/2nd GEARS**	10—50	1	
307434	**SELECTOR FORK, 3rd/4th GEARS**	10—51	1	
307435	**SELECTOR FORK, REVERSE GEAR**	10—52	1	
430475	**Screw,** Taper, securing Selector Forks	10—53	3	
307427	**GEARBOX CHANGE SPEED LEVER**	10—54	1	
432043	**Knob,** on Gear Lever	10—55	1	
433148	**Nut,** securing Knob to Lever	10—56	1	
430486	**Spring,** under Gear Lever Cover Plate	10—59	1	
430478	**Retaining Plate,** for Spring	10—60	1	
307432	**Cover Plate,** for Gear Lever	10—61	1	
UCS.125/3R	**Screw,** Set, securing Cover Plate to Top Cover	10—62	1	
FG.104/X	**Washer,** Spring on Setscrew	10—63	1	
UFB.125/26R	**Bolt,** retaining Gear Lever in Top Cover	10—64	1	
C.8667/1	**Nut,** Nyloc, on Bolt	10—65	1	
307428	**DIPSTICK ASSEMBLY FOR GEARBOX**	10—66	1	
309346	**PRIMARY SHAFT**	11—67	1	
430503	**Bearing Shield,** for Primary Shaft	11—68	1	

Part No.	Description	Plate No.	No. per Unit	Remarks
431693	**BEARING FOR PRIMARY SHAFT (MS.12/KR.V3)**	11—69	1	
431694	**Snap Ring,** on Bearing	11—70	1	
430487	**Washer,** behind Bearing	11—71	1	
422515/14	**Circlip,** retaining Bearing	11—72	1	
431249	**BEARING, PILOT, FOR MAINSHAFT (B.1412)**	11—73	1	
C.22299	**MAINSHAFT**	11—74	1	
430497	**Washer,** at front of Mainshaft Bearing	11—75	1	
431693	**BEARING, FRONT, FOR MAINSHAFT (MS.12/KR.V3)**	11—76	1	
431694	**Snap Ring,** on Bearing	11—77	1	
430487	**Washer,** at rear of Mainshaft Bearing	11—78	1	
422515/14	**Circlip,** retaining Bearing	11—79	1	
430509	**Washer,** locating Mainshaft Rear Bearing	11—80	1	
42315/06	**BEARING, REAR, FOR MAINSHAFT (LS.11)**	11—81	1	
299732	**FLANGE, COUPLING MAINSHAFT TO PROPELLER SHAFT UNIVERSAL JOINT**	11—82	1	
426572	**Washer,** under Flange Nut	11—83	1	
UFN.362/L	**Nut,** Slotted, securing Flange to Mainshaft	11—84	1	
L.104/8U	**Split Pin,** retaining Nut		1	
308143	**SYNCHRONISING SLEEVE, 3rd/4th GEARS**	11—85	1	
425292/07	**Ball,** for 3rd/4th synchronising	11—86	3	
430504	**Spring,** tensioning Ball	11—87	3	
430499	**Shim,** for Spring	11—88	As req'd	
308144	**Outer Synchronising Sleeve**	11—89	1	
309186	**Synchronising Cup,** for 3rd/4th Gears	11—90	2	
309185	**3rd SPEED GEAR ON MAINSHAFT**	11—91	1	
307431	**Bush,** in 3rd Speed Gear	11—92	1	
430500	**Washer,** locating 3rd Speed Gear	11—93	1	
431435	**Spring Ring,** retaining Gear on Mainshaft	11—94	1	

GEARBOX (SYNCHROMESH)
STANDARD TRANSMISSION ONLY

Part No.	Description	Plate No.	No. per Unit	Remarks
309184	**2nd SPEED GEAR, ON MAINSHAFT**	11—95	1	
307429	**Bush,** in 2nd Speed Gear	11—96	1	
C.22298	**Washer,** locating 2nd Speed Gear	11—97	1	
309186	**Synchronising Cup,** for 2nd Speed Gear	11—98	1	
308139	**1st SPEED GEAR SLEEVE**	11—99	1	
430498	**Plunger,** for 1st Gear Interlock	11—100	1	
425292/07	**Ball,** for 1st Gear Interlock	11—101	1	
425292/07	**Ball,** for 2nd Gear Synchro	11—102	3	
430504	**Spring,** for 2nd Gear Synchro Ball	11—103	3	
430499	**Shim,** for 2nd Gear Synchro Ball Spring	11—104	As req'd	
308140	**1st SPEED GEAR, ON MAINSHAFT**	11—105	1	
308145	**COUNTERSHAFT**	11—106	1	
430472	**Retainer,** for Needle Roller on Countershaft	11—107	4	
431695	**Needle Roller,** on Countershaft	11—108	48	
430488	**Thrust Washer,** (·062″ thick) at front of Countershaft	11—109	1	ALTERNATIVES
430489	**Thrust Washer,** (·067″ thick) at front of Countershaft	11—109	1	
430490	**Thrust Washer,** (·072″ thick) at front of Countershaft	11—109	1	
430491	**Thrust Washer,** (·077″ thick) at front of Countershaft	11—109	1	
430492	**Thrust Washer,** (·082″ thick) at front of Countershaft	11—109	1	
430493	**Thrust Washer,** at rear of Countershaft	11—110	1	
308149	**CONSTANT MESH GEAR, ON COUNTERSHAFT**	11—111	1	
430464	**Distance Washer,** between 3rd and Constant Mesh Gears	11—112	1	
308148	**3rd SPEED GEAR, ON COUNTERSHAFT**	11—113	1	
308147	**2nd SPEED GEAR, ON COUNTERSHAFT**	11—114	1	
308146	**1st SPEED GEAR, ON COUNTERSHAFT**	11—115	1	
307422	**REVERSE GEAR SPINDLE**	11—116	1	
430494	**Screw,** locating Reverse Spindle and Countershaft	11—117	1	
FG.105/X	**Washer,** Spring, on Screw	11—118	1	
432135	**REVERSE GEAR ASSEMBLY**	11—119	1	
432592	**Bush,** in Reverse Gear	11—120	2	

GEARBOX (SYNCHROMESH)
STANDARD TRANSMISSION ONLY

Part No.	Description	Plate No.	No. per Unit	Remarks
307422	**REVERSE GEAR FORK**	11—121	1	
307437	**Shaft,** for Reverse Gear Fork	11—122	1	
430511	**Bush,** in Fork, for Shaft	11—123	1	
430512	**Screw,** in Shaft	11—124	1	
430513	**Locknut,** on Screw	11—125	1	
308018	**SPEEDOMETER DRIVE PINION (14 TEETH)**	11—126	1	
431412	**Thrust Washer,** for Speedometer Pinion	11—127	1	
431411	**SPEEDOMETER PINION HOUSING ASSEMBLY**		1	
431409	**Housing,** complete with Bush	11—128	1	
431697	**Oil Seal,** between Housing and Adaptor (025—062)	11—129	1	
431410	**Adaptor,** for Pinion Housing	11—130	1	
422524/01	**Mills Pin,** securing Adaptor		1	
431698	**'O' Ring,** in Pinion Housing	11—131	1	
430519	**Pin,** locating Pinion Housing	11—132	1	
FG.105/X	**Washer,** Spring, on Pin	11—133	1	
UCB.131/11R	**Bolt,** securing Gearbox Assembly to Cylinder Block		7	
UFB.131/10R	**Bolt,** securing Gearbox to Support Plate		4	
UFN.131/L	**Nut,** on Bolts		4	
FG.105/X	**Washer,** Spring, on Bolts and under Nuts		11	
136200	**BRACKET, SUPPORTING GEARBOX**		1	
307497	**RUBBER MOUNTING, FOR GEARBOX**		2	
C.8667/2	**Nut,** securing Rubber Mounting to Support Bracket		2	
UCS.131/5R	**Screw,** Set, securing Rubber Mounting to Gearbox Mounting Bracket		4	
C.8667/2	**Nut,** on Setscrews		4	

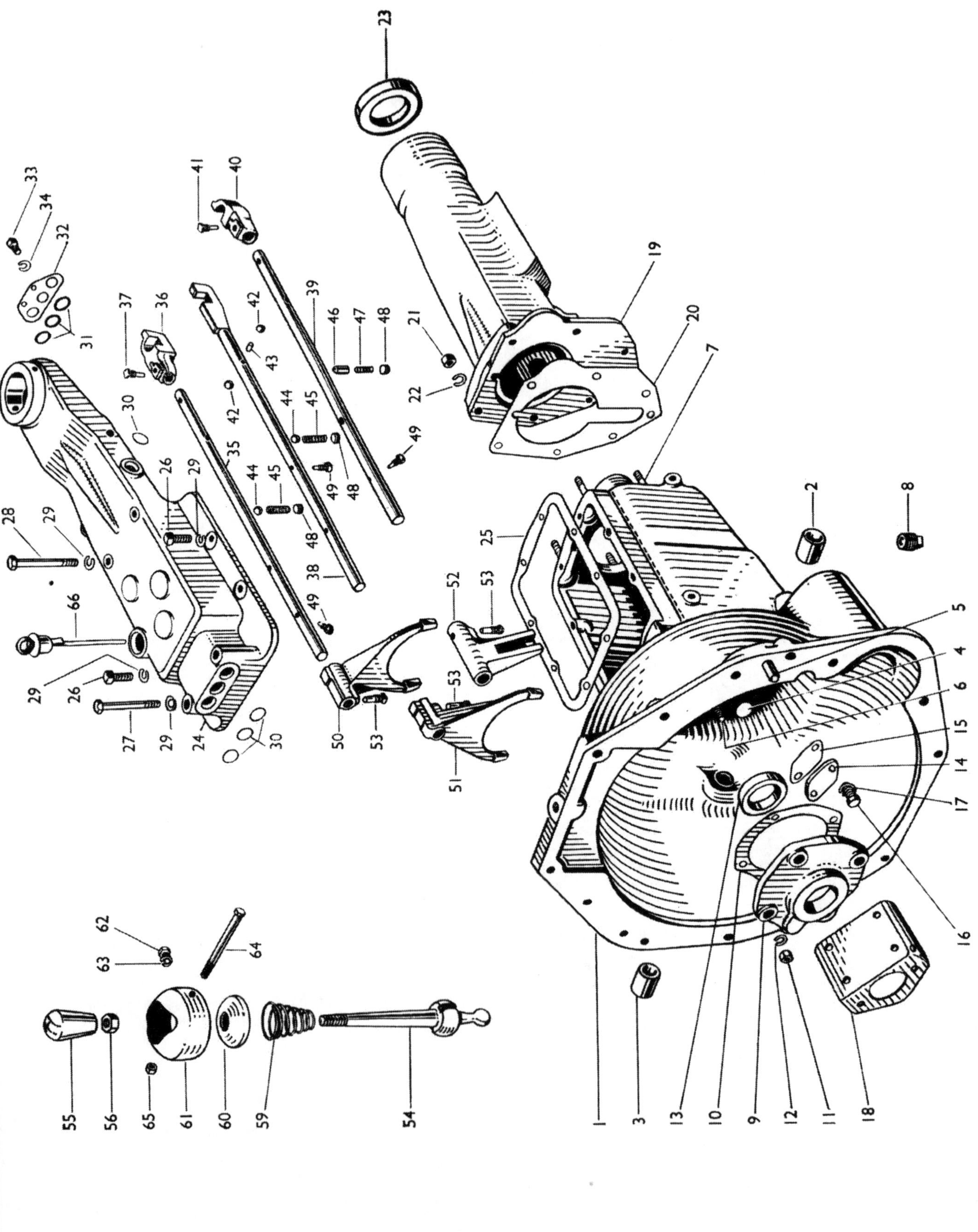

PLATE 10

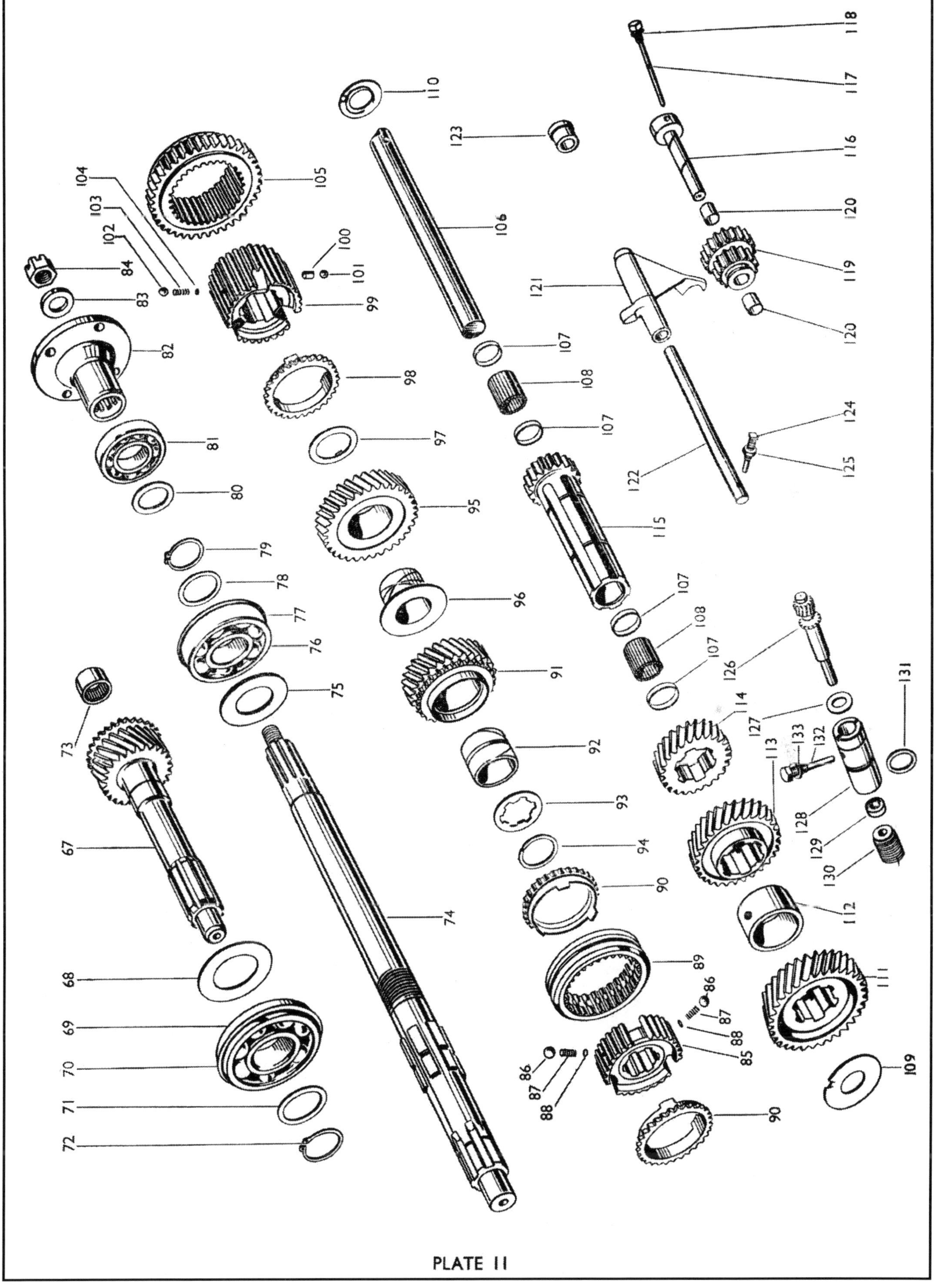
67
68
69
70
71
72
73
74
75
76
77
78
79
80
81
82
83
84
85
86
87
88
89
90
91
92
93
94
95
96
97
98
99
100
101
102
103
104
105
106
107
108
109
110
111
112
113
114
115
116
117
118
119
120
121
122
123
124
125
126
127
128
129
130
131
132
133

PLATE 11

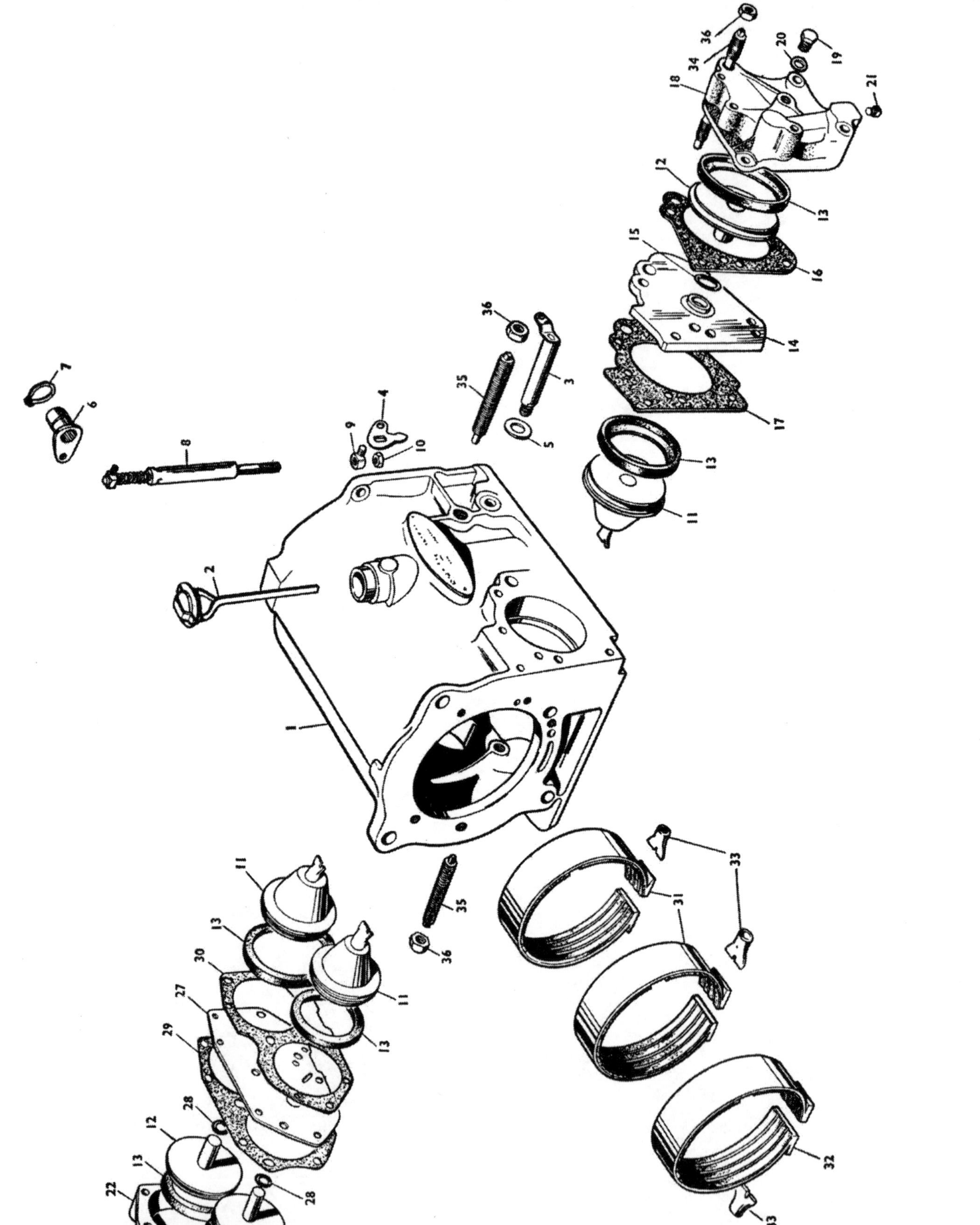

PLATE 12

Part No.	Description	Plate No.	No. per Unit	Remarks
C.21771	**AUTOMATIC TRANSMISSION ASSEMBLY (MODEL DG) (J.20-0294) LESS CONVERTER**		1	
C.21772	**CONVERTER ASSEMBLY (FC.15304)**		1	
8333	**Drain Plug,** for Converter (T.90012)		1	
8334	**Sealing Washer,** on Plug (T.90014)		1	
3701	**TRANSMISSION CASE ASSEMBLY (20-504-A)**	12—1	1	
3702	**Stud,** for Transmission Case (J20-506)		2	
7361	**Oil Seal,** for Selector Control Shaft (J.20-536/A)		1	
6761	**Plug** (Allen Head), under Transmission Case, for Pressure Take-off (JGM.444571)		2	
3740	**DIPSTICK ASSEMBLY (J.20-591)**	12—2	1	
10084	**SELECTOR CONTROL SHAFT ASSEMBLY (J.20-53213)**	12—3	1	
6679	**Selector Lever,** inside Transmission Case (J.20-537/A)	12—4	1	
5455	**Washer,** $\frac{1}{2}$" Special, on Selector Shaft (J.20-538/B)	12—5	1	
6755	**Washer,** $\frac{5}{16}$" Plain, on Selector Shaft (JGM.103340)		1	
5908	**Washer,** $\frac{5}{16}$" lock, on Selector Shaft (JGM.103320)		1	
7002	**Nut,** $\frac{5}{16}$" securing Selector Lever to Control Shaft (JGM.103025)		1	
3750	**TOGGLE ARM LEVER (J.20-630)**	12—6	1	
3751	**Ring,** retaining Toggle Arm Lever in Transmission Case (J.20-633)	12—7	1	
3755	**ROD ASSEMBLY, ACTUATING PARKING BRAKE, FROM SELECTOR CONTROL SHAFT TO TOGGLE ARM LEVER (J.20-6441)**	12—8	1	
6772	**Swivel,** at bottom of Rod (J.20-645)	12—9	1	
6753	**Nut** (J.20-219701)	12—10	1	
3749	**INNER SERVO PISTON ASSEMBLY, $3\frac{1}{4}$" DIAMETER (J.20-622)**	12—11	3	
5463	**OUTER SERVO PISTON ASSEMBLY, $3\frac{1}{4}$" DIAMETER (J.20-605/A)**	12—12	2	
3703	**Seal,** for Inner and Outer Servo Pistons (J.20-512)	12—13	5	

AUTOMATIC TRANSMISSION UNIT

Part No.	Description	Plate No.	No. per Unit	Remarks
5449	**PLATE ASSEMBLY FOR REVERSE SERVO CYLINDER (J.20-519/B)**	12—14	1	
3625	**'O' Ring,** $\frac{1}{2}'' \times \frac{11}{16}'' \times \frac{3}{32}''$ on Reverse Cylinder Plate (J.20-120)	12—15	1	
5450	**Gasket,** for Front of Plate (J.20-523/A)	12—16	1	
5447	**Gasket,** for Rear of Plate (J.20-518/A)	12—17	1	
6828	**REVERSE SERVO CYLINDER (J.20-517/D)**	12—18	1	
8317	**Bolt,** $\frac{5}{16}'' \times 1''$, securing Servo Cylinder to Case (JGM.179818)		1	
8434	**Bolt,** $\frac{5}{16}'' \times 2\frac{3}{4}''$, securing Servo Cylinder to Case (JGM.179829)		3	
8433	**Bolt,** $\frac{5}{16}'' \times 2''$, securing Servo Cylinder to Case (JGM.179826)		2	
5908	**Washer,** $\frac{5}{16}''$ Spring, on Bolts (JGM.103320)		6	
5702	**Plug,** for Pressure Testing on Reverse Gear (J.20-1103)	12—19	1	
3742	**Washer** (Copper) on Plug (J.20-594)	12—20	1	
6761	**Plug,** for Testing Pressure in Converter (JGM.444571)	12—21	1	
7362	**SERVO CYLINDER FOR LOW AND FORWARD SPEEDS (J.20-525/C)**	12—22	1	
8317	**Bolt,** $\frac{5}{16}'' \times 1''$, securing Servo Cylinder to Transmission Case (JGM.179818)		4	
8433	**Bolt,** $\frac{5}{16}'' \times 2''$, securing Servo Cylinder to Transmission Case (JGM.179826)		5	
5908	**Washer,** Spring, on Bolts (JGM.103320)		9	
5702	**Plug,** in Servo Cylinder, for testing pressures on Low and Forward Bands (J.20-1103)	12—23	2	
3742	**Washer,** Copper, on Plugs (J.20-594)	12—24	2	
5444	**PISTON ASSEMBLY FOR LOW LOCK-UP BAND (J.20-513/A)**	12—25	1	
3706	**Seal,** for Piston (J.20-516)	12—26	1	
7363	**PLATE ASSEMBLY FOR LOW BAND SERVO CYLINDER (J.20-5271)**	12—27	1	
3625	**'O' Ring,** $\frac{1}{2}'' \times \frac{11}{16}'' \times \frac{3}{32}''$ between Plate and Piston (J.20-120)	12—28	2	
5452	**Gasket,** between Plate and Servo Cylinder (J.20-526/A)	12—29	1	
6684	**Gasket,** between Plate and Transmission Case (J.20-5291)	12—30	1	
3738	**BRAKE BAND ASSEMBLY FOR LOW AND FORWARD GEARS (J.20-581/A) (COMPLETE WITH LININGS)**	12—31	2	

Part No.	Description	Plate No.	No. per Unit	Remarks
6869	**BRAKE BAND ASSEMBLY FOR REVERSE GEAR (J.20-588/A) (COMPLETE WITH LININGS)**	12—32	1	
3705	**Strut,** for Operation of Brake Bands (J.20-514)	12—33	3	
7528	**Screw,** for adjusting Low Band (J.20-5242)	12—34	1	
5453	**Screw,** for adjusting Forward and Reverse Band (J.20-531/A)	12—35	2	
5910	**Nut,** locking Adjusting Screws (JGM.122295)	12—36	3	
5406	**FRONT PUMP AND COLLECTOR RING ASSEMBLY (J.20-105/A)**	13—1	1	
3622	**Collector Ring Assembly** (J.20-108/B)		1	
3627	**Front Pump Assembly** (J.20-149)		1	
3624	**'O' Ring,** between Collector Ring and Pump Body ($5\frac{1}{2}$" × $5\frac{3}{4}$" × $\frac{1}{8}$") (J.20-119)	13—2	1	
3625	**'O' Ring,** on Centre of Collector Ring ($\frac{1}{2}$" × $\frac{11}{16}$" × $\frac{3}{32}$") (J.20-120)	13—3	1	
3622	**COLLECTOR RING ASSEMBLY (J.20-108/B)**	13—4	1	
3626	**Gasket,** behind Collector Ring (J.20-121)	13—5	1	
3627	**FRONT PUMP BODY ASSEMBLY COMPLETE WITH GEARS (J.20-149)**	13—6	1	
3623	**Oil Seal,** at Front End of Pump (J.20-116)		1	
3810	**Bolt,** $\frac{1}{4}$" × $1\frac{3}{4}$", securing Front Pump to Collector Ring (J.20-1102)		1	
8426	**Washer,** $\frac{1}{4}$" Spring, on Bolt (JGM.103319)		1	
8430	**Bolt,** $\frac{5}{16}$" × $\frac{7}{8}$" securing Front Pump to Collector Ring (JGM.179817)		2	
8432	**Bolt,** $\frac{5}{16}$" × $1\frac{1}{2}$" securing Front Pump to Collector Ring (JGM.179822)		2	
3677	**Thrust Washer,** Controlling End Float on Front Pump and Main Shaft (J.20-388)	13—7	1	CHOOSE THRUST WASHER OF DESIRED THICKNESS TO MAINTAIN CORRECT END FLOAT
3678	**Thrust Washer,** Controlling End Float on Front Pump and Mainshaft (J.20-388/A)	13—7	1	
3679	**Thrust Washer,** Controlling End Float on Front Pump and Mainshaft (J.20-388/B)	13—7	1	
3680	**Thrust Washer,** Controlling End Float on Front Pump and Mainshaft (J.20-388/C)	13—7	1	
8429	**Bolt,** $\frac{5}{16}$" × $\frac{3}{4}$" securing Collector Ring to Transmission Case (JGM.179816)		4	
8431	**Bolt,** $\frac{5}{16}$" × $1\frac{1}{4}$" securing Collector Ring to Transmission Case (JGM.179820)		3	
5908	**Washer,** $\frac{5}{16}$" Spring, on Bolt (JGM.103320)		7	
7881	**MAINSHAFT ASSEMBLY COMPLETE WITH GEAR TRAIN (J.20-3019)**		1	
C.6919	**Tube,** in end of Mainshaft, for Oil Transfer (J.20-392)	13—8	1	

AUTOMATIC TRANSMISSION UNIT

Part No.	Description	Plate No.	No. per Unit	Remarks
3631	**MAINSHAFT ASSEMBLY LESS GEAR TRAIN (J.20-305)**	13—9	1	
3695	**Piston Ring,** 1" × 7/64", on Mainshaft, for Front Planetary Carrier (J.20-413)	13—10	3	
5705	**REAR PLANETARY CARRIER ASSEMBLY (J.20-3152)**	13—11	1	
3699	**Washer,** for Rear Planetary Carrier (J.20-423)	13—12	1	
3637	**Thrust Washer,** for Sun Gear (J.20-321)	13—13	1	
3640	**Thrust Washer,** between Ring Gear and Planetary Carrier (J.20-325)	13—14	1	
3638	**Retainer,** securing Planetary Gears to Mainshaft (J.20-322)	13—15	1	
3691	**Piston Ring,** 1 5/8" × 7/64", in Groove on Retainer (J.20-406)	13—16	1	
3639	**Snap Ring,** securing Retainer (J.20-323)	13—17	1	
5706	**REAR RING GEAR ASSEMBLY (J.20-3262)**	13—18	1	
3689	**'O' Ring,** 1 1/8" × 1 3/4" × 1/8", inside Ring Gear (J.20-401)	13—19	1	
3642	**Piston Ring,** 2 3/8" × 7/64" at rear of Ring Gear (J.20-329/A)	13—20	1	
3690	**'O' Ring,** 5/8" × 1 3/4" × 1/16", at rear of Ring Gear (J.20-402)	13—21	1	
3656	**Snap Ring,** on Splines of Rear Ring Gear (J.20-352)	13—22	1	
5710	**REAR SUN GEAR ASSEMBLY (J.20-3693)**	13—23	1	
3670	**Retainer,** for Rear Sun Gear (J.20-381)	13—24	1	
3669	**Thrust Washer,** for Rear Sun Gear (J.20-380)	13—25	1	
5707	**FRONT SUN GEAR AND LOCK-UP DRUM ASSEMBLY (J.20-3302)**	13—26	1	
3648	**Piston Seal,** 3" diameter (J.20-342)	13—27	1	
3696	**Ring,** for First Clutch Piston (J.20-414)	13—28	1	
3700	**Piston,** complete with Ring, for First Clutch (J.20-424)	13—29	1	
3685	**Plate Assembly,** for First Clutch Retractor (J.20-396)	13—30	1	
6768	**Spacer Plate** (J.20-3451)	13—31	1	
6683	**Plate,** between Friction Discs (J.20-3461)	13—32	4	
6676	**Friction Disc,** for First Clutch (J.20-3472)	13—33	3	
3653	**Retractor Spring,** operating Friction Discs (J.20-348)	13—34	12	
7364	**Retainer Assembly,** for First Clutch Plate (J.20-349/B)	13—35	1	
6764	**Snap Ring,** securing Clutch Plates to Lock-up Drum (J.20-3501)	13—36	1	
5415	**Thrust Washer,** for Front Sun Gear (J.20-351/A)	13—37	1	
5708	**FRONT PLANETARY CARRIER ASSEMBLY (J.20-3532)**		1	
6853	**Planetary Carrier only** (less Brake Drum) (J.20-3561)	13—38	1	
6619	**Reverse Brake Drum** (J.20-360/A)	13—39	1	
3692	**Snap Ring,** retaining Reverse Brake Drum (J.20-409)	13—40	1	
3660	**Thrust Washer,** for Front Planetary Carrier (J.20-362)	13—41	1	

Part No.	Description	Plate No.	No. per Unit	Remarks
5709	**FRONT RING GEAR ASSEMBLY (J.20-3632)**	13—42	1	
3691	**Piston Ring,** 1⅝" diameter × 7/64" for Front Ring Gear (J.20-406)	13—43	3	
5437	**FREE WHEEL UNIT AND LOCK-UP BRAKE DRUM PLATE ASSEMBLY FOR REVERSE GEAR (J.20-395/D)**		1	
5438	**Free Wheel Unit,** Short (J.20-398/D)	13—44	1	
5418	**Lock-Up Plate Assembly** (J.20-367/D)	13—45	1	
3665	**Roller,** positioning Brake Drum Plate to Centre Lock-Up Drum (J.20-373)	13—46	2	
3654	**Snap Ring,** securing Brake Drum Plate in Centre Lock-Up Drum (J.20-350)	13—47	1	
5436	**FREE WHEEL UNIT AND FORWARD BRAKE DRUM ASSEMBLY FOR FORWARD GEARS (J.20-3942)**		1	
7530	**Free Wheel Unit,** Medium (J.20-3891)	13—48	1	
5433	**Forward Brake Drum** (J.20-378/B)	13—49	1	
3667	**Snap Ring,** internal, securing Free Wheel Unit (J.20-376)	13—50	1	
5711	**Spacer,** behind Medium Free Wheel Unit (J.20-3751)	13—51	1	
3694	**Drive Pin,** retaining Spacer (J.20-411)	13—52	1	
3667	**Snap Ring,** securing Spacer (J.20-376)	13—53	1	
3672	**Bearing,** on Mainshaft, behind Free Wheel Unit (J.20-383)	13—54	1	
3673	**Snap Ring,** on Mainshaft Bearing (J.20-384)	13—55	1	
3671	**Spacer,** behind Mainshaft Bearing (J.20-382)	13—56	1	
3693	**Gear,** on Mainshaft, for Parking Brake (J.20-410)	13—57	1	
3674	**Spacer,** between Parking Gear and Bearing (J.20-385)	13—58	1	
3675	**Gear,** on Mainshaft, for Governor Drive (J.20-386)	13—59	1	
3676	**Spacer,** behind Governor Drive Gear (J.20-387)	13—60	1	
C.6867	**UNIVERSAL JOINT COMPANION FLANGE (J.20-093/A)**	13—61	1	
430256	**Bolt,** securing Flange to Propeller Shaft		4	
5439	**Washer,** for Flange (J.20-399/A)	13—62	1	
5926	**Nut,** Self-locking, securing Companion Flange (JGM.272672)	13—63	1	
10088	**VALVE BLOCK ASSEMBLY COMPLETE (J.20-54011)**		1	
10087	**Manifold Plate Assembly** (J.20-5413)		1	
9741	**Base Plate Assembly** (J.20-546/B)	14—1	1	
6565	**Gasket,** between Base Plate and Manifold Plate (J.20-548/A)	14—2	1	
6870	**Relief Valve Body Assembly** (J.20-612/A)		1	
6564	**Gasket,** under Relief Valve Body (J.20-551/A)	14—3	1	

Part No.	Description	Plate No.	No. per Unit	Remarks
	VALVE BLOCK ASSEMBLY (Cont.)			
5915	**Ball,** $\frac{3}{8}$″ dia., for Relief Valve (JGM.147500)	14—4	2	
7368	**Body Assembly,** for Converter Valve (J.20-5602)		1	
7373	**Gasket,** for Converter Valve Body (J.20-5732)	14—5	1	
6851	**Bolt,** $\frac{1}{4}$″ × $2\frac{3}{8}$″, for Assembly of Valve Block (J.20-1104)		8	
8424	**Washer,** $\frac{1}{4}$″ Spring, on Bolts (J.20-5942)		8	
6757	**Bolt,** $\frac{5}{16}$″ × $1\frac{1}{8}$″, securing Valve Block to Transmission Case (JGM.106326)		7	
8423	**Washer,** $\frac{5}{16}$″ Spring, on Bolts (J.20-5941)		7	
10087	**MANIFOLD PLATE ASSEMBLY (J.20-5413)**	14—6	1	
10089	**Spring,** for Selector Valve Detent (J.20-544/A)	14—7	1	
5914	**Ball,** $\frac{1}{4}$″ diameter, for Selector Valve (JGM.147485)	14—8	1	
6870	**RELIEF VALVE BODY ASSEMBLY (J.20-612/A)**	14—9	1	
6680	**Retainer,** for Anti-Bump Valve Plug (J.20-570/A)	14—10	1	
6868	**Stop,** for Main Relief Valve (J.20-567/A)	14—11	1	
6682	**Retainer,** for Accumulator Spring (J.20-667)	14—12	1	
3732	**Retainer,** for Reverse Interlock Spring (J.20-520)	14—13	1	
7368	**CONVERTER VALVE BODY ASSEMBLY (J.20-5602)**	14—14	1	
7369	**Body only,** for Converter Valve (J.20-5612)		1	
6754	**Converter Shuttle Valve,** Direct (J.20-6381)	14—15	1	
6779	**Converter Shuttle Valve,** Reverse (J.20-6414)	14—16	1	
6999	**Sleeve,** for Converter Shuttle Valve (J.20-6431)	14—17	1	
6681	**Retainer,** for Shuttle Valve Sleeve (J.20-617/A)	14—18	1	
6688	**OIL FILTER ASSEMBLY (J.20-5572)**	14—19	1	
6873	**Retainer,** for Oil Filter (J.20-558/B)	14—20	1	
6759	**Screw,** securing Retainer (JGM.113954)		1	
C.739	**Washer,** Shakeproof, on Screw (JGM.114604)		1	
5467	**OIL SUMP (J.20-649/A)**	14—21	1	
3737	**Plug,** in Oil Sump Drain Hole (J.20-580)	14—22	1	
3741	**Washer,** on Plug (J.20-592)	14—23	1	
3736	**Gasket,** between Oil Sump and Transmission Case (J.20-579)	14—24	1	
6757	**Bolt,** $\frac{5}{16}$″ × $\frac{5}{8}$″, securing Oil Sump to Transmission Case (JGM.106324)		14	
5908	**Washer,** $\frac{5}{16}$″ Spring, on Bolts (JGM.103320)		14	
10090	**EXTENSION CASE ASSEMBLY COMPLETE (J.20-70259)**		1	
10091	**Extension Case Assembly** (J.20-7048)	15—1	1	

Part No.	Description	Plate No.	No. per Unit	Remarks
	EXTENSION CASE ASSEMBLY (Cont.)			
3625	**'O' Ring,** on Bush for Governor Adjusting Shaft ($\frac{1}{2}$" × $\frac{11}{16}$" × $\frac{3}{32}$") (J.20-120)	15—2	1	
5469	**Pawl,** second Speed Drive (J.20-719/A)	15—3	1	
3764	**Washer,** for Control Link and Pawl (J.20-720)	15—4	2	
3765	**Rocker Arm,** for Governor Control (J.20-721)	15—5	1	
3767	**Washer,** behind Rocker Arm (J.20-723)		1	
5688	**Stop,** for Governor (J.20-722/A)	15—6	1	
3768	**Distance Piece,** behind Stop Plate (J.20-724)		1	
3769	**Spring,** operating Governor Control Rocker Arm (J.20-725)	15—7	1	
3808	**Shaft and Lever Assembly,** for Governor adjustment (J.20-836)	15—8	1	
3770	**Lever,** Governor Adjusting (J.20-728)	15—9	1	
10085	**Finger,** for Governor Adjusting Lever (J.20-7361)	15—10	2	
3771	**Return Spring,** for Governor Adjusting Lever (J.20-735)	15—11	1	
3795	**Valve and Fork Assembly,** for Governor Control (J.20-790)	15—12	1	
3774	**Detent Piston,** for Governor Control Valve (J.20-744)	15—13	1	
8428	**Screw,** $\frac{1}{4}$" × $\frac{3}{8}$" securing Retaining Plate for Governor Fork (JGM.114354)		1	
8426	**Washer,** $\frac{1}{4}$" Spring, on Screw (JGM.103319)		1	
3796	**Plunger,** in Extension, operating Accelerator Detent (J.20-791)	15—14	1	
3797	**Spring,** tensioning Plungers (J.20-792)	15—15	1	
3798	**Gasket,** between Extension and Transmission Case (J.20-793)	15—16	1	
3762	**Oil Seal,** at end of Extension Case (J.20-716)	15—17	1	
8435	**Bolt,** $\frac{3}{8}$" × 1", securing Extension Case to Transmission Case (JGM.179839)		4	
5700	**Bolt,** $\frac{3}{8}$" × $2\frac{1}{4}$", securing Extension Case to Transmission Case (J.20-1100)		1	
5701	**Bolt,** $\frac{3}{8}$" × 3" securing Extension Case to Transmission Case (J.20-1101)		1	
8427	**Washer,** $\frac{3}{8}$" Spring, on Bolts (JGM.103321)		6	
5715	**DETENT CAM, CONTROL LINK AND PLUNGER ASSEMBLY, OPERATING KICKDOWN CONTROL (J.20-8301)**	15—18	1	
6758	**Key,** Woodruff $\frac{1}{2}$" diameter × $\frac{1}{16}$" positioning Control Shaft to Detent Cam (JGM.106958)	15—19	1	
6889	**Screw, Set** (Allen Head) No. 10 × $\frac{3}{16}$" preventing withdrawal of Shaft from Cam (JGM.140879)	15—20	1	
3771	**Return Spring,** for Detent Cam (J.20-735)	15—11	1	
3793	**Screw,** Set, $\frac{1}{4}$" × $1\frac{1}{4}$", for Kickdown Adjustment (J.20-787)	15—21	1	
6752	**Washer** (Copper), on Setscrew (JGM.446195)	15—22	1	
3794	**Nut,** Stop, for Setscrew (J.20-778)	15—23	1	
5954	**Nut,** Dome, for Setscrew (JGM.441263)	15—24	1	
10092	**GOVERNOR ASSEMBLY (J.20-7751)**	15—25	1	
3788	**Bearing,** on Governor Shaft (J.20-778)		1	
3782	**Washer** (Fibre) on Governor Shaft (J.20-767)		1	
3789	**Snap Ring,** retaining Bearing and Washer on Shaft (J.20-779)		2	
3785	**Shaft,** carrying Governor Assembly (J.20-773/A)	15—26	1	
7372	**Return Spring,** on Governor Shaft (J.20-7848)	15—27	1	

AUTOMATIC TRANSMISSION UNIT

Part No.	Description	Plate No.	No. per Unit	Remarks
3799	**PAWL AND TOGGLE ASSEMBLY OPERATING PARKING DEVICE (J.20-800)**	15—28	1	
7535	**Pivot Pin,** for Pawl (J.20-8081)	15—29	1	
7536	**Shaft,** supporting Parking Brake (J.20-8051)	15—30	1	
3802	**Ring,** retaining Shaft in Extension (J.20-807)	15—31	1	
3805	**INTERLOCK PISTON ASSEMBLY FOR PARKING BRAKE (J.20-810)**	15—32	1	
10037	**Spring,** for Parking Brake Interlock Piston (J.20-8131)	15—33	1	
3783	**HOUSING FOR SPEEDOMETER DRIVE GEAR (J.20-770)**	15—34	1	
3786	**Gasket,** between Housing and Extension (J.20-774)	15—35	1	
3792	**Cover,** for Housing (J.20-786)	15—36	1	
3791	**Gasket,** between Housing and Cover (J.20-785)	15—37	1	
3784	**SPEEDOMETER DRIVE GEAR (J.20-772/A)**	15—39	1	
3775	**REAR PUMP ASSEMBLY (J.20-750)**		1	
5689	**Body only,** for Rear Pump (J.20-751)	15—40	1	
5690	**Driving Gear and Shaft Assembly** (J.20-755)	15—41	1	
5691	**Idler Gear Assembly** (J.20-758)	15—42	1	
5692	**Cover,** for Rear Pump (J.20-762)	15—43	1	
3779	**Gasket** (·002″ thick) under Cover (Coloured clear) (J.70-761)	15—44	1	
3780	**Gasket** (·003″ thick) under Cover (Coloured clear) (J.20-761/A)	15—44	1	
3778	**Gasket** (·005″ thick) under Cover (Coloured Grey) (J.20-761/B)	15—44	1	**SELECTIVE RANGE FOR CLEARANCE BETWEEN COVER AND GEARS** (3779, 3780, 3778, 3776, 3777)
3776	**Gasket** (·006″ thick) under Cover (Coloured Opaque Black or Manilla) (J.20-716/C)	15—44	1	
3777	**Gasket** (·008″ thick) under Cover (Coloured Black or Manilla) (J.20-761/D)	15—44	1	
8422	**Bolt,** $\frac{1}{4}$″ × $\frac{7}{8}$″, securing Cover to Pump Body (J.20-1108)		2	
8426	**Washer,** $\frac{1}{4}$″ Spring, on Bolts (JGM.103319)		2	
5693	**Governor Driven Gear** (J.20-765)	15—45	1	
6756	**Pin,** securing Governor Driven Gear (JGM.103723)	15—46	1	
3781	**Gasket,** between Rear Pump and Extension (J.20-766)	15—47	1	
8318	**Bolt,** $\frac{5}{16}$″ × $2\frac{1}{4}$″ securing Rear Pump to Extension (JGM.-179827)		3	
5908	**Washer,** $\frac{5}{16}$″ Spring, on Bolts (JGM.103320)		3	
7341	**Plug,** in Rear Pump Cover (JGM.444732)		1	
431792	**SPEEDOMETER DRIVEN GEAR ASSEMBLY**	15—48	1	
308731	**REAR ENGINE MOUNTING BRACKET ON GEARBOX**		1	
UCS.137/6R	**Screw,** Set, securing Mounting Bracket to Gearbox		4	
FG.106/X	**Washer,** Spring, on Setscrews		4	

Part No.	Description	Plate No.	No. per Unit	Remarks
6767	**COMPLETE KIT OF OIL SEALS FOR USE WHEN OVERHAULING TRANSMISSION UNIT (J.20-097)**		1	
7476	**GASKET SET FOR USE WHEN OVERHAULING TRANSMISSION UNIT**		1	
8782	**REPAIR KIT FOR MULTI-DISC CLUTCH (J.20-0984)**		1	
	Consisting of:—			
6683	**Clutch Plate**		4	
6676	**Friction Disc**		3	
3653	**Retractor Spring**		12	
6764	**Snap Ring**		1	
3648	**Piston Seal**		1	
8783	**THRUST WASHER KIT (J.20-099)**		1	
	Consisting of:—			
3637	**Thrust Washer**		1	
3677	**Thrust Washer**		1	
3678	**Thrust Washer**		1	
3679	**Thrust Washer**		1	
3680	**Thrust Washer**		1	
3640	**Thrust Washer**		1	
3669	**Thrust Washer**		1	
5415	**Thrust Washer**		1	
3660	**Thrust Washer**		1	

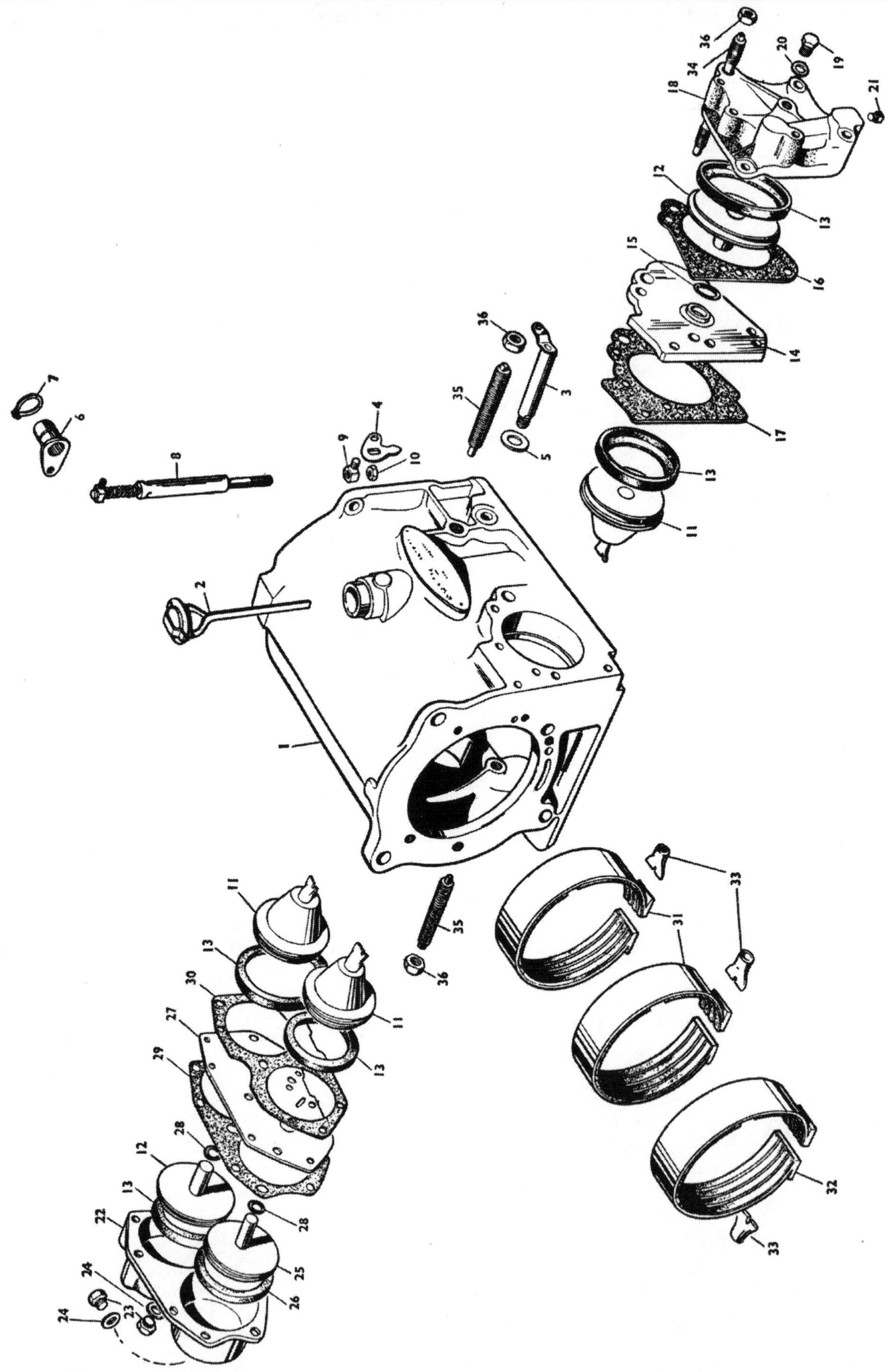
36
34
20
19
18
21
12
13
15
16
14
36
7
6
35
3
4
9
10
5
17
8
13
11
2
1
33
31
11
35
13
30
36
31
11
27
13
29
28
32
12
13
22
28
33
25
24
26
24
23

PLATE 12

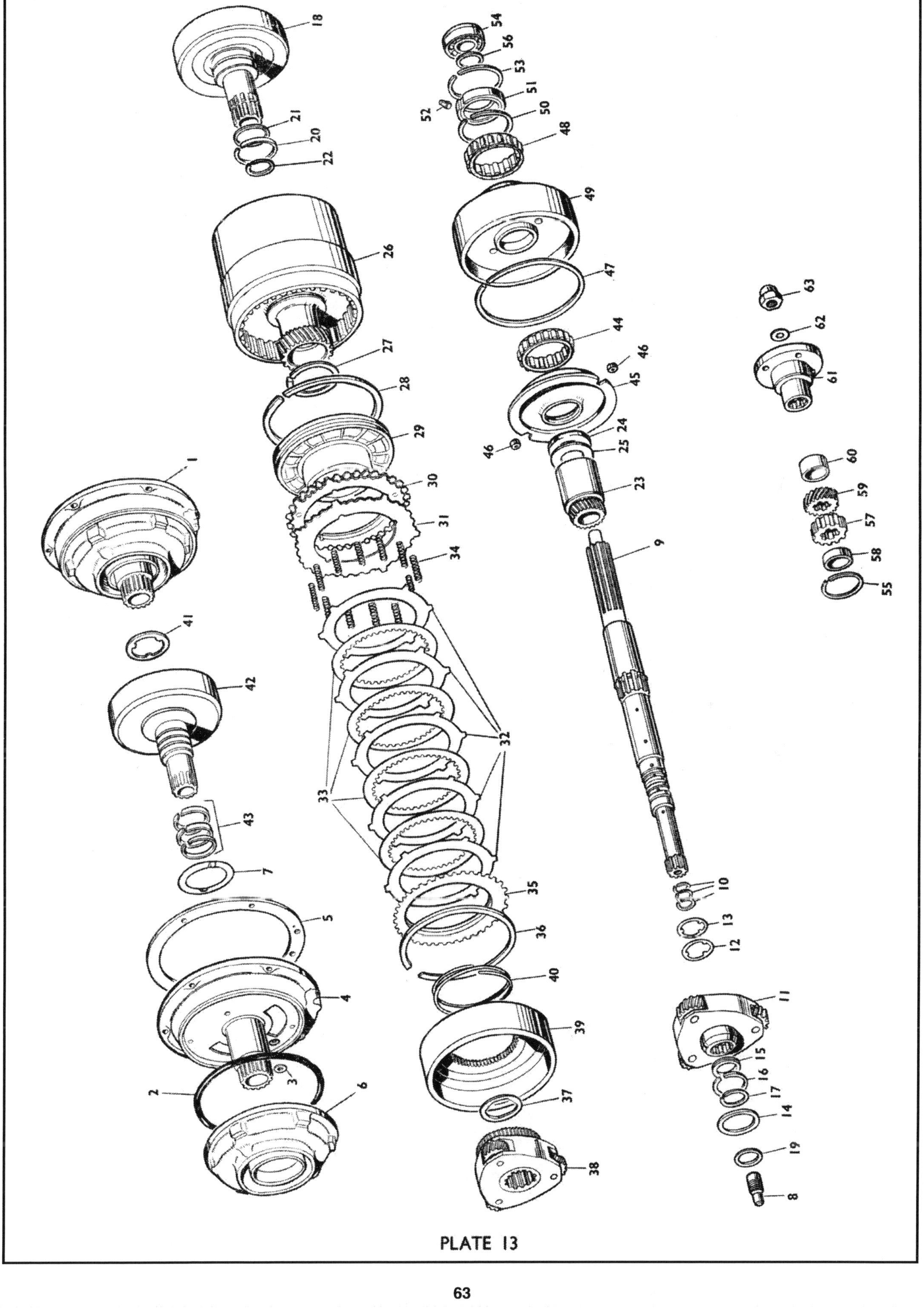
18
21
20
22
54
56
53
51
52
50
48
49
26
47
63
62
44
46
27
28
45
24
61
29
46
25
1
30
23
60
31
59
57
9
34
58
55
41
42
32
33
43
7
10
35
13
5
36
12
40
4
11
39
15
3
16
2
6
17
37
14
19
38
8

PLATE 13

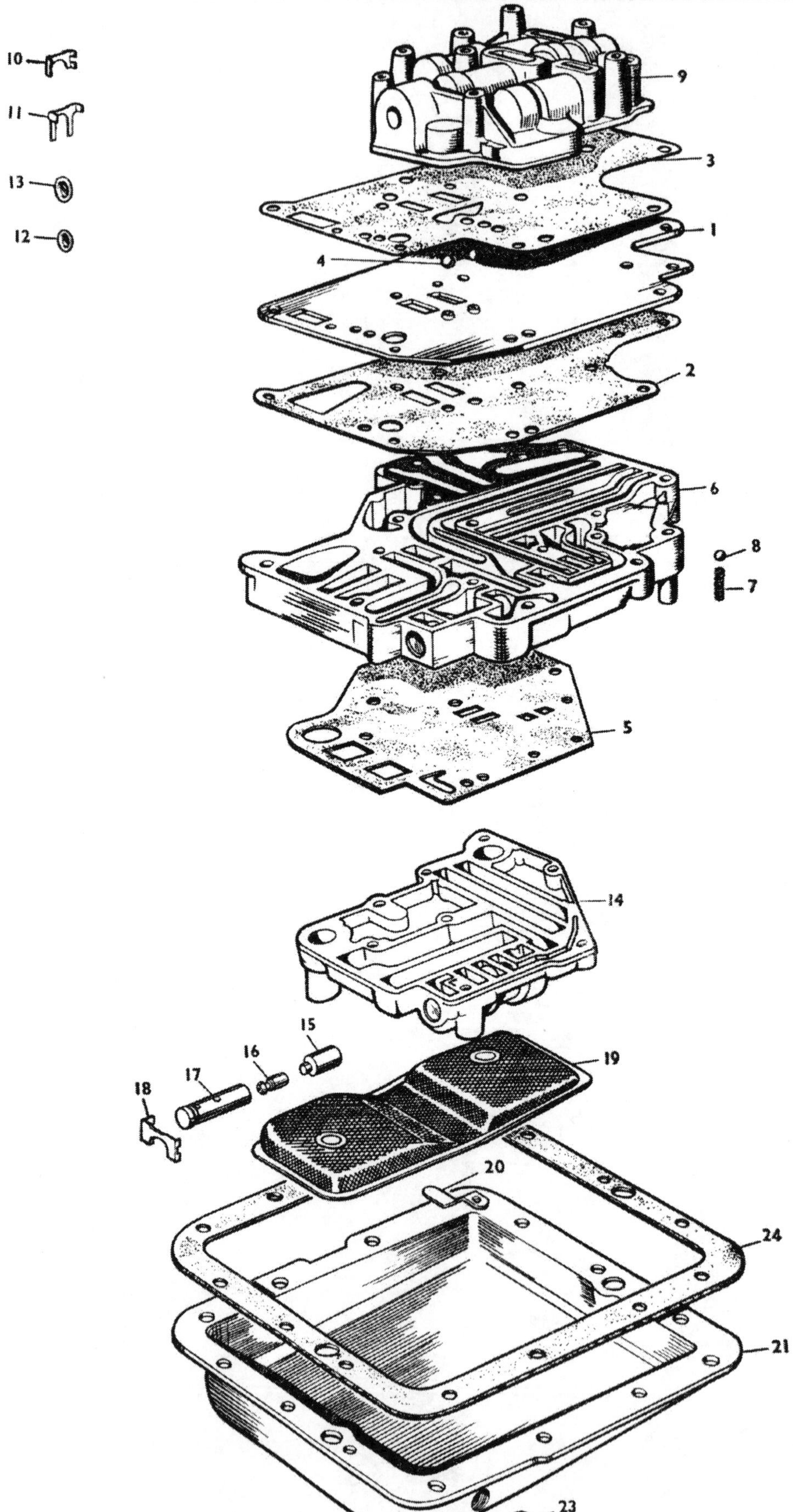

PLATE 14

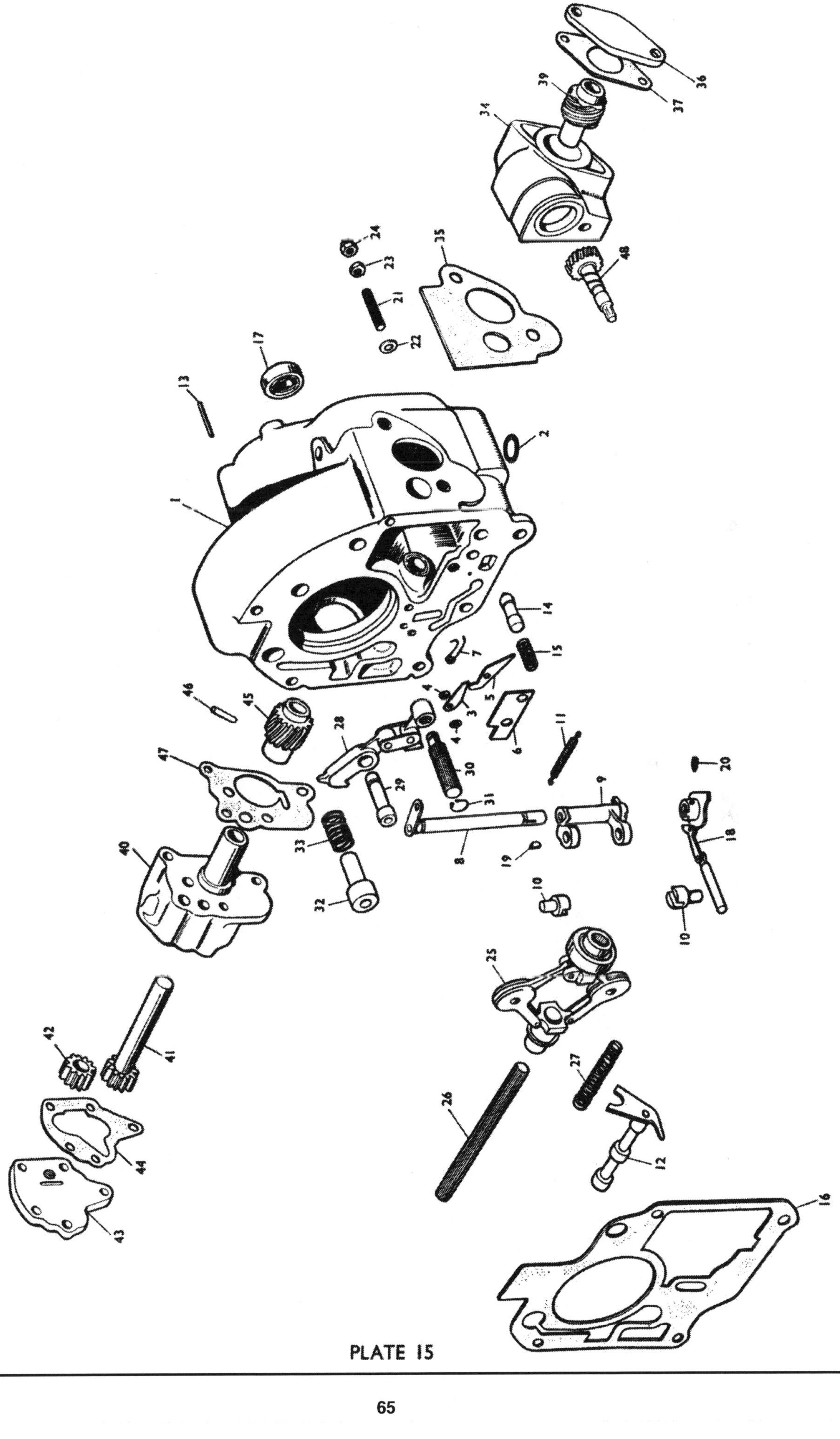
36
37
39
34
24
23
35
48
21
22
17
13
2
1
14
15
7
5
4
3
46
45
28
4
6
11
20
47
30
29
31
9
33
18
8
19
40
10
10
32
25
42
41
27
26
44
12
16
43

PLATE 15

AUTOMATIC TRANSMISSION UNIT

MOUNTING DETAILS

Part No.	Description	Plate No.	No. per Unit	Remarks
136462	**CONVERTER HOUSING ASSEMBLY**	16—1	1	
0608/16	**Bush,** for Converter Housing	16—2	1	
420304	**Dowel,** for Converter Housing	16—3	1	
34007/16	**Stud,** $\frac{7}{16}$" × 2", for Converter Housing	16—4	2	
UFN.143/L	**Nut,** $\frac{7}{16}$" securing Converter Housing to Gearbox		4	
FG.107/X	**Washer,** $\frac{7}{16}$" Spring, on Studs		4	
307193	**Cover Plate,** for Air Outlet	16—5	2	
432864	**Cover Plate,** for Air Inlet	16—6	2	
NS.419/4H	**Screw,** No. 10 × $\frac{1}{2}$", securing Cover		12	
AG.102/X	**Washer,** Spring, on Screws		12	
307391	**STARTER GEAR RING**	16—7	1	
135832	**Mounting Plate,** for Starter Gear Ring	16—8	1	
430205	**Screw** (Long), securing Starter Gear Ring to Mounting Plate		6	
430206	**Screw** (Short), securing Starter Gear Ring to Mounting Plate		2	
FG.106/X	**Washer,** $\frac{3}{8}$" Spring, on Screws		8	
307226	**Spacing Washer,** for Mounting Plate	16—9	2	
430239	**Bolt,** for Crankshaft Flange	16—10	6	
430237	**Lock Plate,** on Bolts	16—11	3	
UCS.131/12R	**Setscrew,** securing Converter Housing and Support Plate to Cylinder Block		7	
UFB.131/11R	**Bolt,** securing Converter Housing to Transmission Support Plate		6	
UFN.131/L	**Nut,** on Bolts		6	
FG.105/X	**Washer,** Spring, on Setscrews and Bolts		13	
136200	**BRACKET, SUPPORTING TRANSMISSION UNIT**	16—12	1	
307497	**RUBBER MOUNTING FOR TRANSMISSION UNIT**	16—13	2	
C.8667/2	**Nut,** securing Rubber Mounting to Support Bracket		2	
UFS.131/7R	**Screw,** Set, securing Rubber Mounting to Mounting Bracket		4	
C.8667/2	**Nut,** on Setscrews		4	

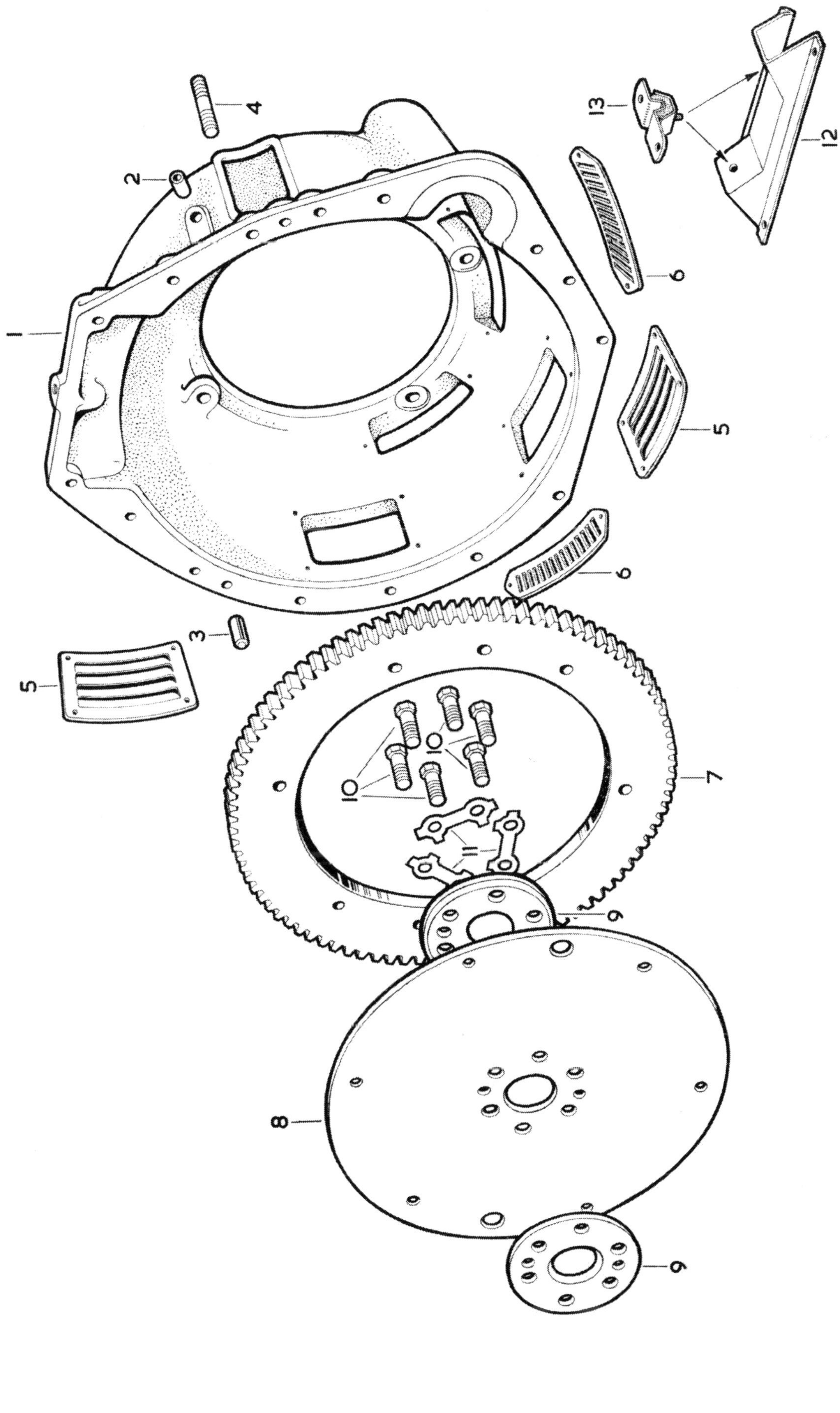

PLATE 16

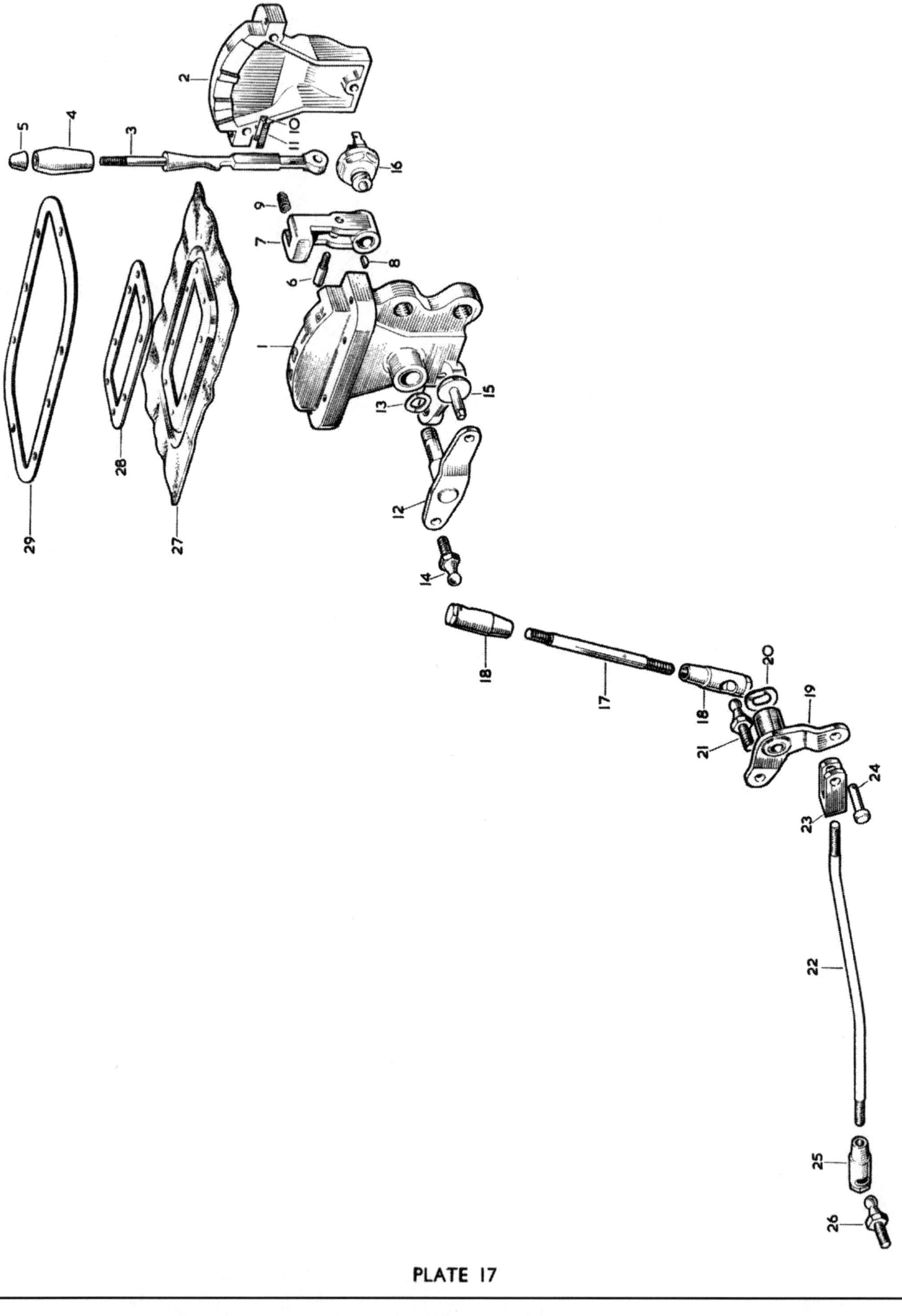
1
2
3
4
5
6
7
8
9
10
11
12
13
14
15
16
17
18
18
19
20
21
22
23
24
25
26
27
28
29

PLATE 17

Part No.	Description	Plate No.	No. per Unit	Remarks
432438	**SELECTOR QUADRANT ASSEMBLY**	17—1	1	
0810/16	**Bush,** in Quadrant		1	
UCS.131/5R	**Screw,** Set, securing Quadrant to Transmission Unit Mounting Bracket		3	
FG.105/X	**Washer,** Spring, on Setscrews		3	
136454	**COVER FOR QUADRANT ASSEMBLY**	17—2	1	
UCS.525/8H	**Screw,** Set, securing Cover to Quadrant		1	
UCS.525/10H	**Screw,** Set, securing Cover to Quadrant		2	
423145	**Lockwasher,** on Setscrews		3	
307192	**SELECTOR HAND LEVER**	17—3	1	
430195	**Knob,** on Selector Lever	17—4	1	
430199	**Nut,** securing Knob to Lever	17—5	1	
430194	**Pivot Pin,** securing Selector Lever to Boss on Operating Lever	17—6	1	
UFN.119/L	**Nut,** securing Pivot Pin		1	
C.723/A	**Washer,** Shakeproof, under Nut		1	
308734	**BOSS ON OPERATING LEVER**	17—7	1	
422510/10	**Key,** locating Boss on Operating Lever	17—8	1	
UFN.137/L	**Nut,** securing Boss to Operating Lever		1	
FG.106/X	**Washer,** Spring, under Nut		1	
430200	**Spring,** tensioning Selector Hand Lever	17—9	1	
430242	**Stop Plate,** for Selector Hand Lever	17—10	1	
430243	**Packing** (Fibre) under Stop Plate	17—11	1	
906049	**Screw,** Set, securing Retaining Plate to Boss		2	
432420	**OPERATING LEVER, IN QUADRANT ASSEMBLY**	17—12	1	
432943	**Wavy Washer,** between Operating Lever and Quadrant	17—13	1	
426768	**Ball Pin,** in Operating Lever for Control Rod	17—14	1	
UFN.125/L	**Nut,** securing Ball Pin		1	
432922	**CONTACT ADJUSTER ON OPERATING LEVER, FOR STARTER SOLENOID CUT-OUT**	17—15	1	
NN.121/L	**Nut,** securing Contact Adjuster to Operating Lever		1	
308374	**SWITCH IN QUADRANT, FOR STARTER SOLENOID CUT-OUT**	17—16	1	
C.730	**Washer,** Shakeproof, on Switch		1	

SELECTOR CONTROLS
(AUTOMATIC TRANSMISSION)

Part No.	Description	Plate No.	No. per Unit	Remarks
422542/6	**CONTROL ROD, BETWEEN OPERATING LEVER AND BELL CRANK LEVER**	17—17	1	
426767	**Ball Socket,** on Control Rod	17—18	2	
UFN.125/L	**Nut,** securing Ball Sockets to Rod		2	
FG.104/X	**Washer,** Spring, under Nuts		2	
432480	**BELL CRANK LEVER ASSEMBLY, ON TRANSMISSION UNIT MOUNTING BRACKET**	17—19	1	
0608/16	**Bush,** on Bell Crank Lever		1	
432942	**Wavy Washer,** between Bell Crank Lever and Bracket	17—20	1	
C.8667/1	**Nut,** securing Bell Crank Lever to Bracket		1	
FW.104/T	**Washer,** Plain, under Nut		1	
426768	**Ball Pin,** on Bell Crank Lever for Control Rod	17—21	1	
UFN.125/L	**Nut,** securing Ball Pin to Bell Crank Lever		1	
FG.104/X	**Washer,** Spring under Nut		1	
308736	**CONTROL ROD, FROM BELL CRANK LEVER TO SELECTOR LEVER ON TRANSMISSION UNIT**	17—22	1	
432436	**Fork End,** connecting Control Rod to Bell Crank Lever	17—23	1	
422531/03	**Clevis Pin,** securing Fork End to Bell Crank Lever	17—24	1	
L.102/4U	**Split Pin,** retaining Clevis Pin		1	
FW.104/T	**Washer,** Plain, behind Split Pin		1	
426767	**Ball Socket,** connecting Control Rod to Selector Lever on Transmission Unit	17—25	1	
UFN.125/L	**Nut,** securing Fork End and Ball Socket to Rod		2	
426768	**BALL PIN, ON TRANSMISSION UNIT SELECTOR LEVER**	17—26	1	
UFN.125/L	**Nut,** securing Ball Pin		1	
FG.104/X	**Washer,** Spring, under Nut		1	
308765	**LEATHER GAITER, FOR QUADRANT ASSEMBLY**	17—27	1	
432444	**Fixing Plate** (Inner) securing Gaiter to Quadrant	17—28	1	
426934/22	**Screw,** securing Fixing Plate to Quadrant		8	
AG.104/X	**Washer,** Spring, on Screws		8	
308763	**Fixing Plate** (Outer) securing Gaiter	17—29	1	
425285/17	**Screw,** securing Fixing Plate to Gearbox Cover		10	

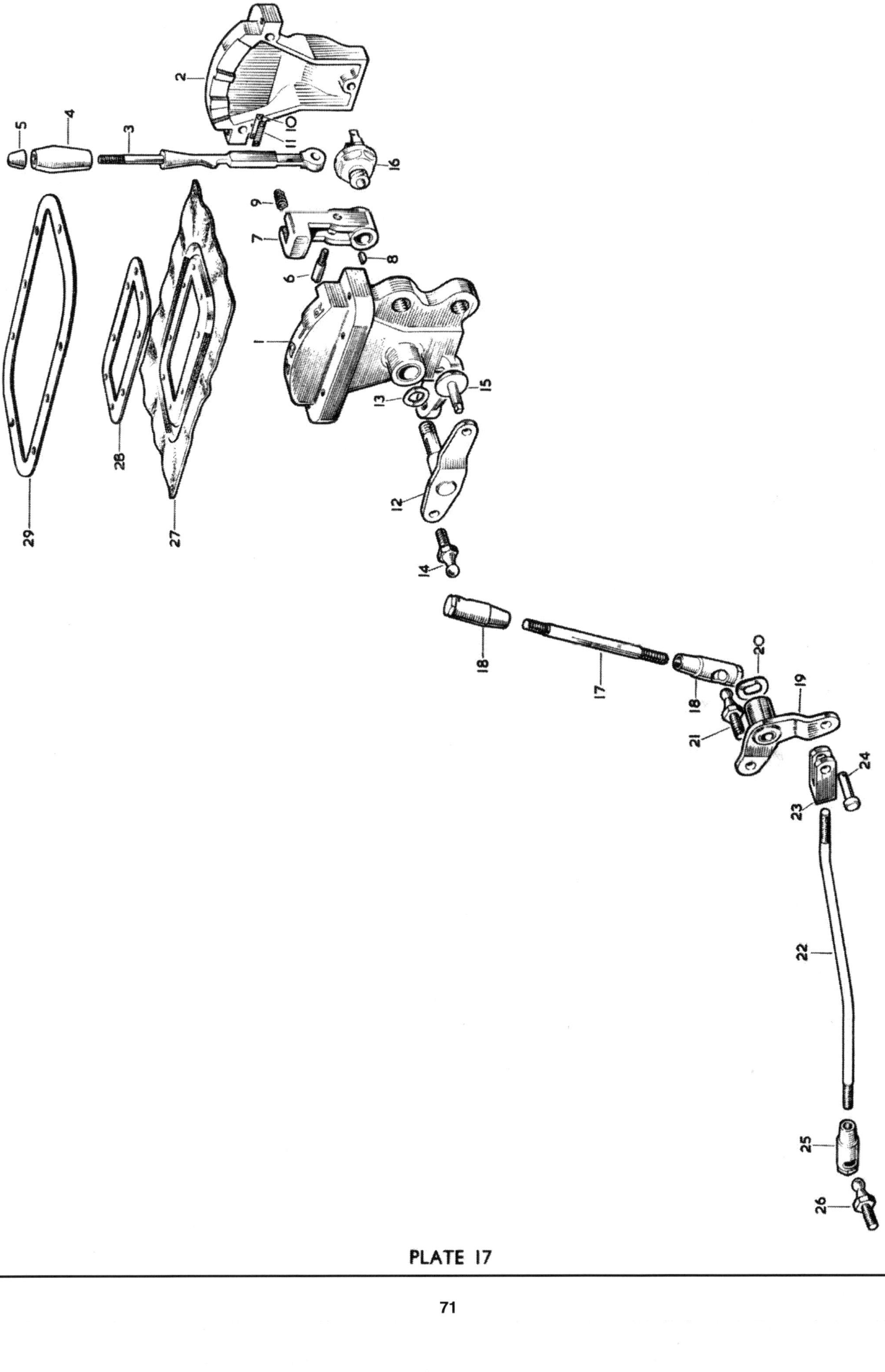
1
2
3
4
5
6
7
8
9
10
11
12
13
14
15
16
17
18
18
19
20
21
22
23
24
25
26
27
28
29

PLATE 17

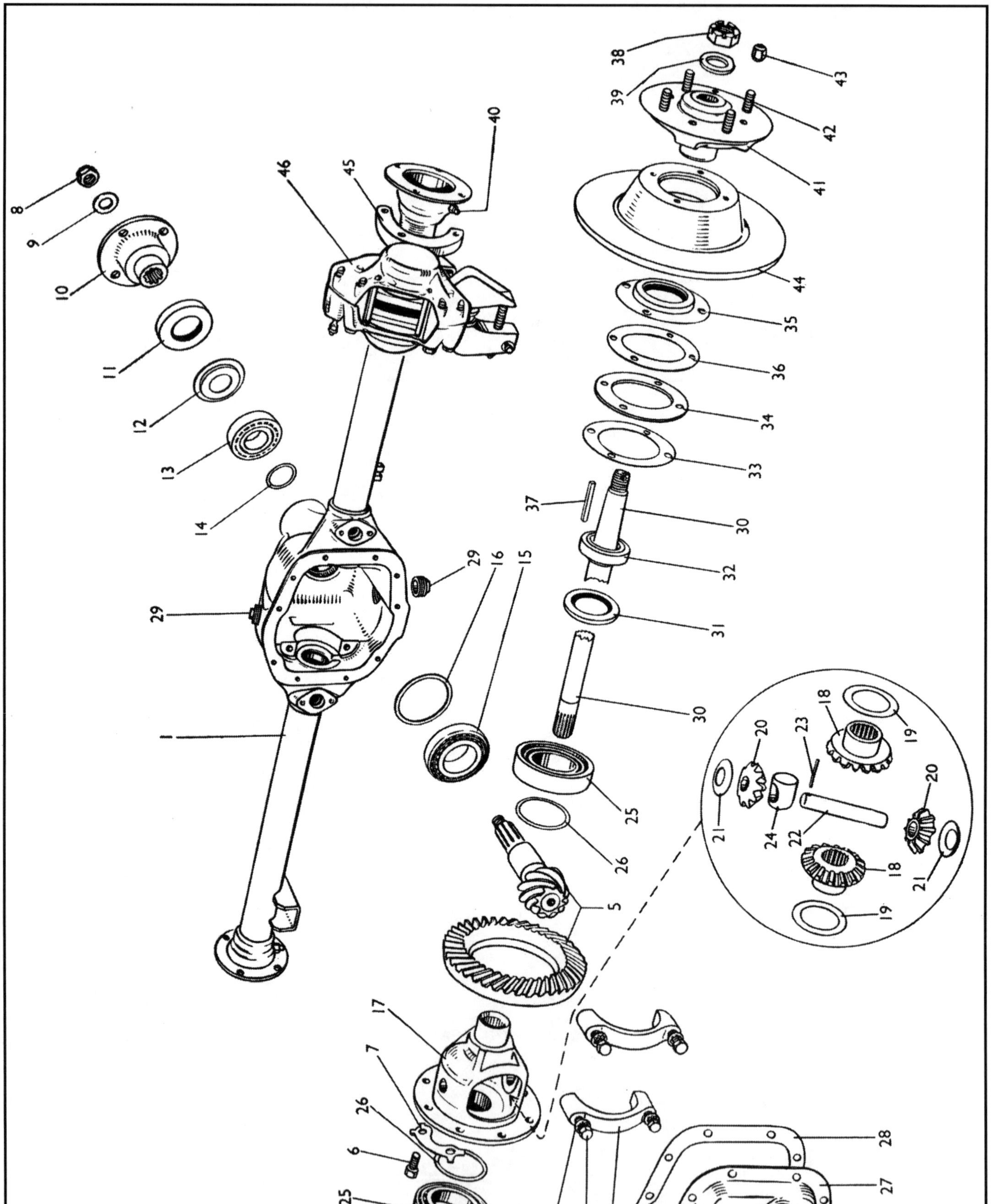

8
9
10
11
12
13
14
29
1
46
45
40
16
15
29
38
39
43
42
41
44
35
36
34
33
37
30
32
31
30
25
26
5
20
23
18
19
21
24
22
20
18
21
19
17
7
26
6
25
30
4
3
2
28
27
29

PLATE 19

Part No.	Description	Plate No.	No. per Unit	Remarks
302150	**PROPELLER SHAFT ASSEMBLY**	18—1	1	**FOR USE ON CARS FITTED WITH STANDARD TRANSMISSION ONLY**
12	**Flange Yoke** (K2/2/239)	18—2	2	
13	**Sleeve Yoke** (K2/3/68)	18—3	1	
26	**Dust Cap,** for Sleeve Yoke (3/14/39)	18—4	1	
28	**Washer,** (Cork) in Dust Cap (K3/16/43)	18—5	1	
27	**Washer,** (Steel) in Dust Cap (K3/15/33)	18—6	1	
8406	**Grease Nipple,** on Sleeve Yoke (94/GB.2204)	18—7	1	
46	**Journal Assembly,** connecting Flange Yokes to Propeller Shaft (K.5/L.4)	18—8	2	
248	**Grease Nipple,** in Journals (94/GB.1598)	18—9	2	
309289	**PROPELLER SHAFT ASSEMBLY**		1	**FOR USE ON CARS FITTED WITH AUTOMATIC TRANSMISSION ONLY**
	Individual items are as for Propeller Shaft Assembly 302150			
430256	**Bolt,** securing Propeller Shaft to Gearbox and Rear Axle		8	**STANDARD TRANSMISSION ONLY**
C.8667/3	**Nut,** on Bolts		8	
430256	**Bolt,** securing Propeller Shaft to Rear Axle		4	**AUTOMATIC TRANSMISSION ONLY**
C.8667/3	**Nut,** on Bolts at Rear Axle		4	
C.8667/3	**Nut,** on Bolts, in Transmission Unit Flange		4	

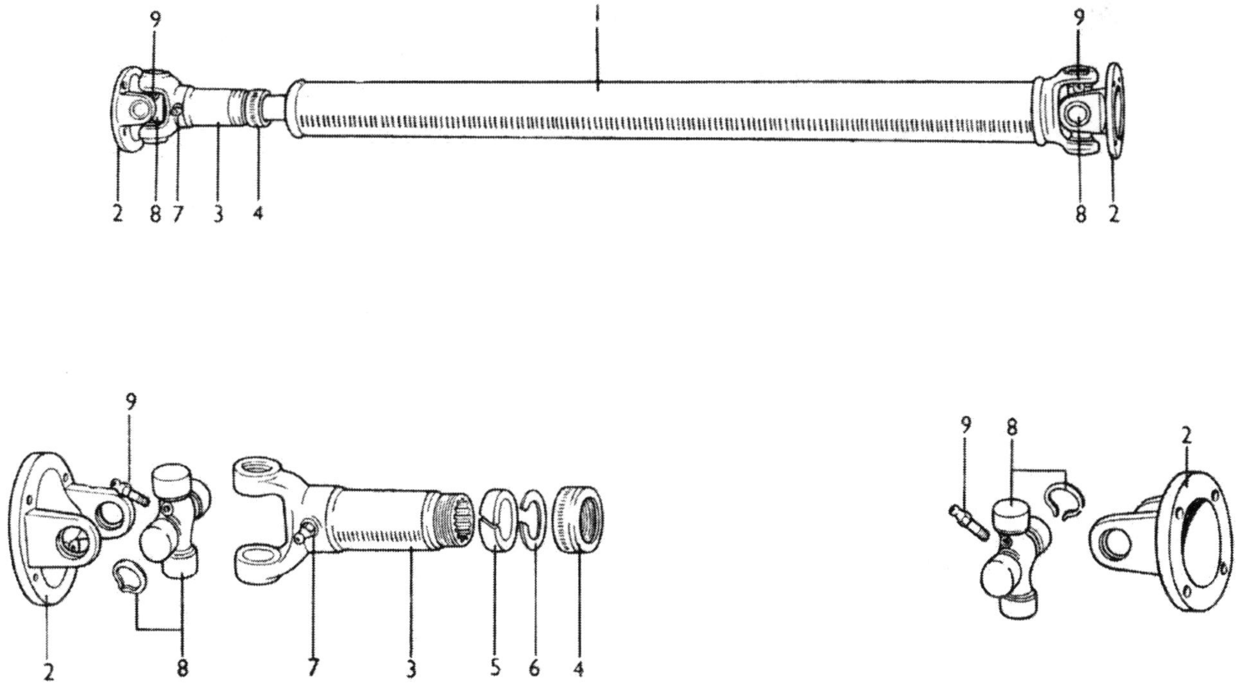

PLATE 18

REAR AXLE

Part No.	Description	Plate No.	No. per Unit	Remarks
26223	**REAR AXLE ASSEMBLY COMPLETE (RATIO 3·58:1) (7HA-001-10)**		1	**1st CONDITION**
26223/01	**Carrier and Tube Assembly** (7HA-101/6)	19—1	1	
135220/02	**Differential Bearing Cap** (6HA-011/1)	19—2	2	
135220/40	**Screw,** Set, securing Bearing Caps (6B-NC.32)	19—3	4	
6124	**Lockwasher,** on Setscrews (6LW-11)	19—4	4	
26223/02	**Crown Wheel and Pinion** (Ratio 3·58:1) (7HA-105/7)	19—5	1	
26223/03	**Screw,** Set, securing Crown Wheel to Differential Case (7HA-075)	19—6	8	
26223/04	**Lockplate,** on Setscrews (7HA-074)	19—7	4	**NO LONGER REQUIRED**
135220/07	**Nut,** securing Companion Flange to Drive Pinion (10LN-NF.8)	19—8	1	
135220/08	**Washer,** under Nut (10W-19)	19—9	1	
26223/05	**Companion Flange Assembly** (7HA-102/4)	19—10	1	
26223/06	**Oil Seal,** on Pinion (7HA-019/1)	19—11	1	
135220/13	**Oil Slinger,** on Pinion (7HA-036)	19—12	1	
26223/07	**Roller Bearing,** at outer end of Pinion (7HA-022)	19—13	1	
135220/15	**Shim,** outer, adjusting Pinion (·003″ thick) (6HA-039)	19—14	As req'd	
135220/16	**Shim,** outer, adjusting Pinion (·005″ thick) (6HA-040)	19—14	As req'd	**USE SHIM OF DESIRED THICKNESS**
135220/17	**Shim,** outer, adjusting Pinion (·010″ thick) (6HA-041)	19—14	As req'd	
135220/18	**Shim,** outer, adjusting Pinion (·030″ thick) (6HA-042)	19—14	As req'd	
3844	**Roller Bearing,** at inner end of Pinion (2HA-023)	19—15	1	
3860	**Shim,** inner, adjusting Pinion (·003″ thick) (2HA-043)	19—16	As req'd	
3861	**Shim,** inner, adjusting Pinion (·005″ thick) (2HA-044)	19—16	As req'd	**USE SHIM OF DESIRED THICKNESS**
3862	**Shim,** inner, adjusting Pinion (·010″ thick) (2HA-045)	19—16	As req'd	
26223/08	**Case,** Differential Assembly (7HA-082/3)		1	
135220/30	**Roller Bearing,** on Differential Case (6HA-024/1)	19—25	2	
135220/31	**Shim,** adjusting Differential (·003″ thick) (6HA-046/1)	19—26	As req'd	
135220/32	**Shim,** adjusting Differential (·005″ thick) (6HA-047/1)	19—26	As req'd	**USE SHIM OF DESIRED THICKNESS**
135220/33	**Shim,** adjusting Differential (·010″ thick) (6HA-048/1)	19—26	As req'd	
135220/34	**Shim,** adjusting Differential (·030″ thick) (6HA-049/1)	19—26	As req'd	
26223/11	**Cover,** for Carrier (7HA-010/1)	19—27	1	
26223/12	**Gasket,** for Cover (7HA-026)	19—28	1	
CS.131/5D	**Screw,** Set, securing Cover to Carrier (5B-NC.10)		10	
47	**Lockplate,** on Setscrews (5LW-95)		10	
3823	**Plug,** for Filler and Drain Apertures (HA-059)	19—29	3	
26223/13	**Axle Shaft** (7HA-005/2)	19—30	2	
26280	**REAR AXLE COMPLETE (RATIO 3·58:1) (7HA-001-30)**		1	**2nd CONDITION**
9188	**Carrier and Tube Assembly** (7HA-101/15)		1	
9196	**Case,** Differential Assembly (7HA-082/7)		1	
9199	**Roller Bearing,** on Differential Case (7HA-024)		2	
3863	**Shim,** adjusting Differential (·003″ thick) (2HA-046)		As req'd	
3864	**Shim,** adjusting Differential (·005″ thick) (2HA-047)		As req'd	**USE SHIM OF DESIRED THICKNESS**
3865	**Shim,** adjusting Differential (·010″ thick) (2HA-048)		As req'd	
3866	**Shim,** adjusting Differential (·030″ thick) (2HA-049)		As req'd	
9189	**Crown Wheel and Pinion** (Ratio 3·58:1) (7HA-105/11)		1	
9192	**Roller Bearing,** at inner end of Pinion (7HA-023/1)		1	
9193	**Shim,** inner, adjusting Pinion (·003″ thick) (7HA-043)		As req'd	
9194	**Shim,** inner, adjusting Pinion (·005″ thick) (7HA-044)		As req'd	**USE SHIM OF DESIRED THICKNESS**
9195	**Shim,** inner, adjusting Pinion (·010″ thick) (7HA-045)		As req'd	
9191	**Roller Bearing,** at outer end of Pinion (7HA-022/1)		1	
9190	**Oil Seal,** on Pinion (6HA-091/1)		1	

All other items are as for Rear Axle 26223

Part No.	Description	Plate No.	No. per Unit	Remarks
26223/08	**DIFFERENTIAL CASE ASSEMBLY (7HA-082/3)**		1	**1st CONDITION**
26223/09	**Case only** (7HA-006/3)	19—17	1	
4132	**Gear,** Side, for Differential (3HA-007/9)	19—18	2	
3912	**Thrust Washer,** for Side Gears (3HA-038)	19—19	2	
4133	**Gear,** Differential Pinion Mate (3HA-008/9)	19—20	2	
4135	**Thrust Washer,** for Pinion Mate Gears (3HA-037/2)	19—21	2	
26223/10	**Shaft,** for Pinion Mate Gears (7HA-012)	19—22	1	
3838	**Pin,** locking Pinion Mate Gears to Differential Case (2HA-013)	19—23	1	
3909	**Spacer,** on Shaft (3HA-033/1)	19—24	1	
9196	**DIFFERENTIAL CASE ASSEMBLY (7HA-082/7)**		1	**2nd CONDITION**
9197	**Case only** (7HA-006/5)		1	
9198	**Pin,** locking Pinion Mate Gears to Differential Case (7HA-013)		1	
26223/23	**Spacer,** on Shaft (7HA-033)		1	
	All other items are as for Differential Case Assembly 26223/08			
26223/13	**REAR AXLE SHAFT (7HA-005/2)**	19—30	2	
4144	**Oil Seal,** in ends of Carrier Tube (3HA-027/3)	19—31	2	
7993	**Roller Bearing,** for Hubs (7HA-025)	19—32	2	
4145	**Shim,** adjusting Hub Bearing (·003″ thick) (HA-050/2)	19—33	As req'd	**USE SHIM OF DESIRED THICKNESS**
4146	**Shim,** adjusting Hub Bearing (·005″ thick) (HA-051/2)	19—33	As req'd	
4147	**Shim,** adjusting Hub Bearing (·010″ thick) (HA-052/2)	19—33	As req'd	
4148	**Shim,** adjusting Hub Bearing (·030″ thick) (HA-053/2)	19—33	As req'd	
26223/16	**Retainer Plate,** for Hub Bearings (7HA-035/2)	19—34	2	
3039	**Oil Seal,** for Rear Hubs (HA-057/1)	19—35	2	
4149	**Gasket,** between Oil Seal and Retainer Plate (HA-031/4)	19—36	2	
7757	**Bolt,** securing Retainer Plate to Axle Tube (6B-NF.18B)		2	
26223/22	**Bolt,** securing Retainer Plate and Brake Caliper Support Bracket to Axle Tube (6B-NF.26)		6	
26223/19	**Nut,** on Bolts (6LN-NF.6B)		8	
134502/03	**Washer,** Plain, under Nuts (6W-12)		8	
C.11835	**Key,** locating Hubs on Shafts (3HA-029/1)	19—37	2	
4141	**Slotted Nut,** on Axle Shafts (14SN-NF.10)	19—38	2	
L.104/14U	**Split Pin,** retaining Nuts (8SP-13)		2	
4141	**Washer,** Plain, under Nuts (14W-26)	19—39	2	
422521/29	**Grease Nipple,** for Hub Bearings (6HA-056)	19—40	2	
26223/21	**Shim,** (·030″ thick) for Caliper Support Bracket (7HA-053/1)		As req'd	
307996	**REAR HUB ASSEMBLY**	19—41	2	
430649	**Stud,** in Hubs, for Road Wheels	19—42	8	
430650	**Nut,** securing Road Wheels to Hubs	19—43	8	
308124	**REAR BRAKE DISC**	19—44	2	
431685	**Screw,** Set, securing Brake Discs to Hubs		8	
431492	**Lockplate,** on Setscrews		4	

Part No.	Description	Plate No.	No. per Unit	Remarks
308118	**SUPPORT BRACKET FOR REAR BRAKE CALIPERS**	19—45	2	
136203	**REAR BRAKE CALIPER R.H.**	19—46	1	
432668	**Bolt,** securing Caliper to Support Bracket		4	
431155	**Lockplate,** on Bolts		2	
431907/01	**Shim** (·005" thick) adjusting Caliper		As req'd	SELECTIVE THICKNESS
431907/02	**Shim** (·020" thick) adjusting Caliper		As req'd	SELECTIVE THICKNESS

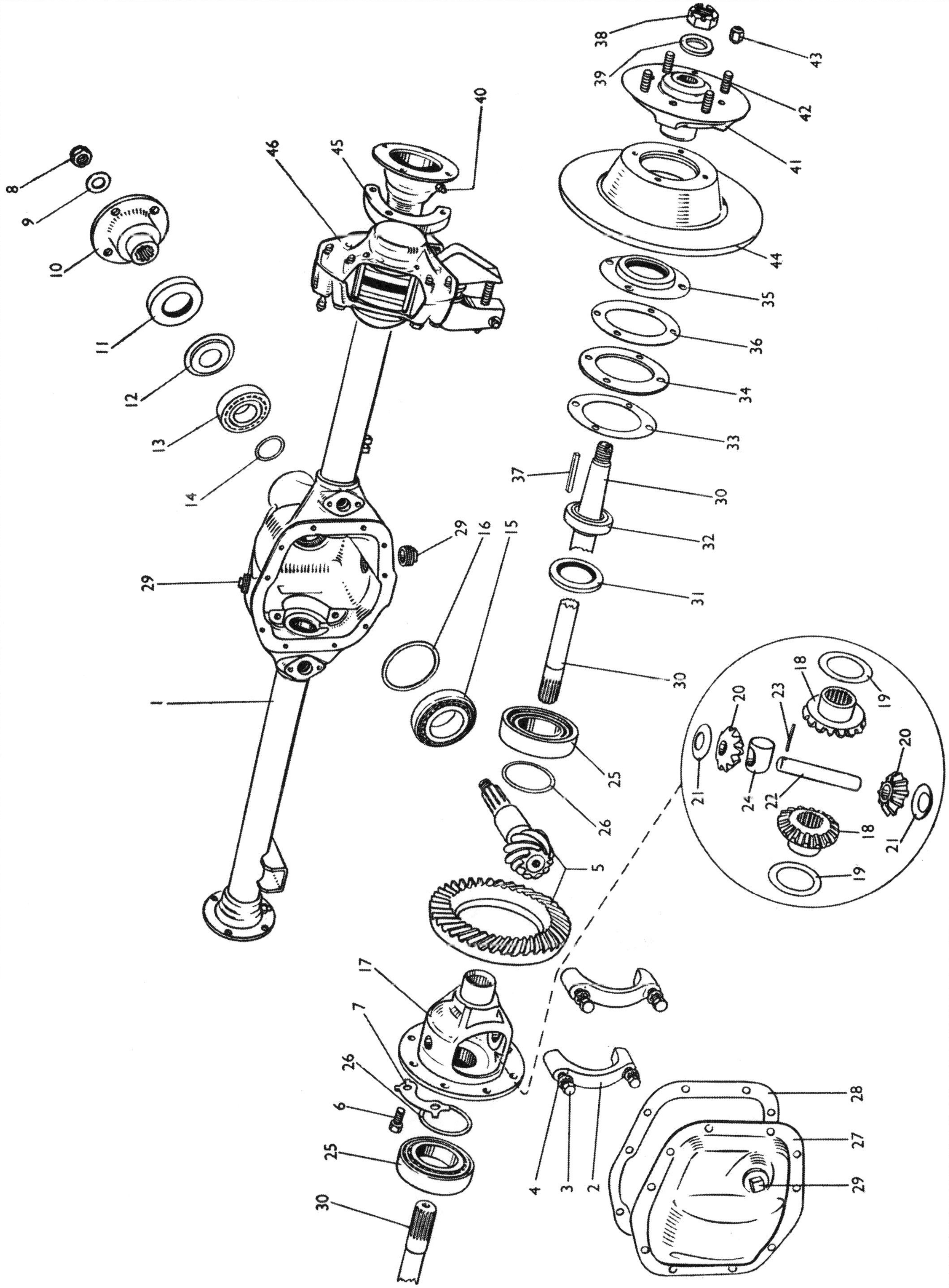

8
9
10
11
12
13
14
29
1
46
45
40
38
39
43
42
41
44
35
36
34
33
37
30
32
31
29
16
15
30
25
26
5
18
19
20
23
21
24
22
20
18
21
19
17
7
26
6
25
30
4
3
2
28
27
29

PLATE 19

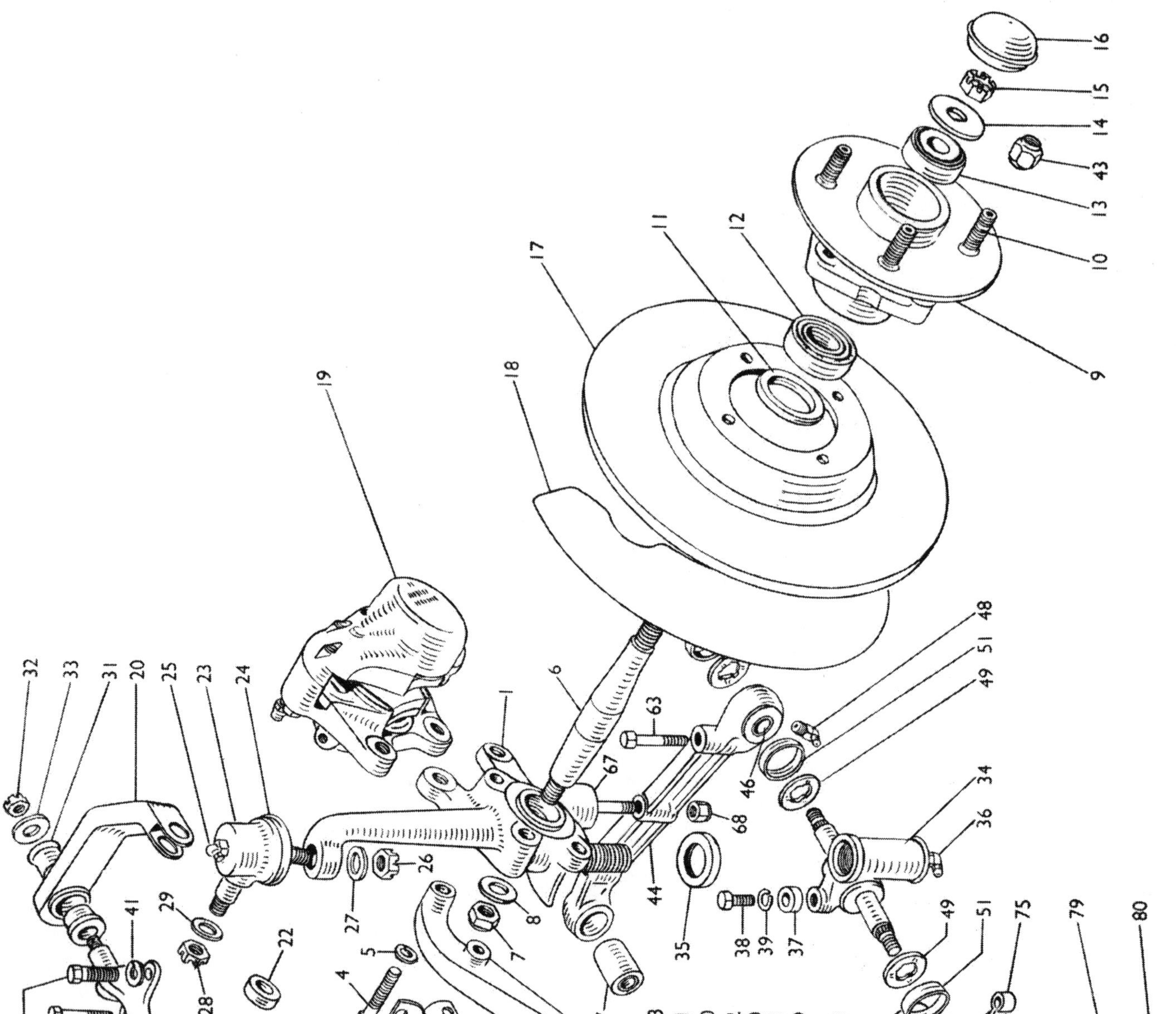

PLATE 20

Part No.	Description	Plate No.	No. per Unit	Remarks
136194	**R.H. FRONT SUSPENSION UNIT ASSEMBLY**		1	
136195	**L.H. FRONT SUSPENSION UNIT ASSEMBLY**		1	
136221	**R.H. STUB AXLE CARRIER**		1	
136222	**L.H. STUB AXLE CARRIER**	20—1	1	
307531	**R.H. STEERING ARM**		1	
307532	**L.H.STEERING ARM**	20—2	1	
308933	**Brake Hose Abutment,** on Steering Arm Bolts	20—3	2	
431557	**Bolt,** securing Steering Arms to Stub Axle Carriers	20—4	4	
FG.106/X	**Washer,** Spring, on Bolts	20—5	4	
308004	**STUB AXLE SHAFT**	20—6	2	
308004/01	**Nut,** Nyloc, on inner end of Shaft	20—7	2	
FW.109/T	**Washer,** Plain, under Nuts	20—8	2	
308174	**FRONT HUB ASSEMBLY**	20—9	2	
431383	**Stud,** for Road Wheels	20—10	8	
431566	**Oil Seal Assembly,** for Front Hubs	20—11	2	
431570	**Bearing,** Inner, for Front Hubs (1988/1922)	20—12	2	
C.15352	**Bearing,** Outer, for Front Hubs (LM.11949/LM.11910)	20—13	2	
C.3400	**'D' Washer,** on Stub Axle Shaft	20—14	2	
NN.756/L	**Slotted Nut,** securing Front Hubs to Stub Axle Shaft	20—15	2	
L.104/8U	**Split Pin,** retaining Nuts		2	
C.3039	**HUB CAP**	20—16	2	
136252	**FRONT BRAKE DISC (64325226)**	20—17	2	
431386	**Bolt,** securing Discs to Hubs		8	
FG.106/X	**Washer,** Spring, on Bolts		8	
308172	**DUST COVER ASSEMBLY FOR R.H. BRAKE DISC**		1	

FRONT SUSPENSION

Part No.	Description	Plate No.	No. per Unit	Remarks
308173	**DUST COVER ASSEMBLY FOR L.H. BRAKE DISC**	20—18	1	
UFS.137/5R	**Screw,** Set, securing Dust Covers to Stub Axle Carriers		2	
FG.106/X	**Washer,** Spring, on Setscrews		2	
FW.106/T	**Washer,** Plain, under Spring Washers		2	
136158	**R.H. FRONT BRAKE CALIPER (64032501 CV)**		1	
136157	**L.H. FRONT BRAKE CALIPER (64032500 CV)**	20—19	1	
431560	**Shim,** (·010″ thick) adjusting Calipers to Stub Axle Carriers		As req'd	
431559	**Bolt,** securing Calipers and Dust Covers to Stub Axle Carriers		4	
431558	**Tab Washer,** on Bolts		2	
308179	**UPPER WISHBONE ARM (RIGHT-HAND FRONT, LEFT-HAND REAR)**	20—20	2	
308180	**UPPER WISHBONE ARM (LEFT-HAND FRONT, RIGHT-HAND REAR)**	20—21	2	
431581	**Distance Piece,** in fork end of front Wishbone Arms	20—22	2	
308183	**BALL JOINT ASSEMBLY FOR L.H. UPPER WISHBONE**	20—23	1	
C.3020	**Gaiter.** for Ball Joint	20—24	1	
431579	**Grease Nipple,** for Ball Joint	20—25	1	
308182	**BALL JOINT ASSEMBLY FOR R.H. UPPER WISHBONE**		1	
	Individual items are as for Ball Joint Assembly 308183			
431589	**Slotted Nut,** securing Ball Joints to Stub Axle Carriers	20—26	2	
L.104/10U	**Split Pin,** retaining Slotted Nuts		2	
FW.108/T	**Washer,** Plain, under Nuts	20—27	2	
431581	**Slotted Nut,** securing Ball Joints to Upper Wishbones	20—28	2	
L.104/8U	**Split Pin,** retaining Slotted Nuts		2	
FW.109/T	**Washer,** Plain, under Nuts	20—29	2	
308181	**INNER FULCRUM FOR UPPER WISHBONE**	20—30	2	
431573	**Rubber Bush,** in Fulcrums	20—31	2	
431590	**Slotted Nut,** securing Upper Wishbones to Fulcrums	20—32	4	
L.103/7U	**Split Pin,** retaining Slotted Nuts		4	
431574	**Washer,** Plain, under Nuts	20—33	4	

Part No.	Description	Plate No.	No. per Unit	Remarks
308187	**LOWER TRUNNION ASSEMBLY**	20—34	2	
431584	**Grease Seal,** between Lower Trunnions and Stub Axle Carriers	20—35	2	
431579	**Grease Nipple,** for Lower Trunnion	20—36	2	
431586	**LOCK-STOP, ON LOWER TRUNNION**	20—37	2	
UFS.131/6R	**Screw,** Set, securing Lock-stop to Lower Trunnion	20—38	2	
FG.105/X	**Washer,** Spring, on Setscrews	20—39	2	
UFS.137/9R	**Screw,** Set, securing Inner Fulcrum to Chassis	20—40	4	
FG.106/X	**Washer,** Spring, on Setscrews	20—41	4	
UFB.137/14R	**Bolt,** securing Inner Fulcrum to Chassis	20—42	4	
C.8667/3	**Nut,** Nyloc, on Bolts		4	
430650	**Nut,** securing Road Wheels to Hubs	20—43	8	
431416	**LOWER WISHBONE ARM ASSEMBLY (RIGHT-HAND FRONT, LEFT-HAND REAR)**	20—44	2	
431415	**LOWER WISHBONE ARM ASSEMBLY (RIGHT-HAND REAR, LEFT-HAND FRONT)**	20—45	2	
431588	**Bush,** in outer end of Wishbone Arm	20—46	4	
431414	**Bush,** in inner end of Wishbone Arm (13/1343)	20—47	4	
431579	**Grease Nipple,** for Wishbone Arm	20—48	4	
431389	**Thrust Washer,** for Lower Trunnion Shackle Pins	20—49	8	
431391	**Lockwasher,** under Nut on Trunnion Shackle Pins	20—50	4	
431390	**Grease Seal,** between Lower Wishbones and Trunnions	20—51	8	
431590	**Slotted Nut,** securing Lower Wishbone Arms to Trunnions	20—52	4	
L.103/7U	**Split Pin,** retaining Slotted Nuts		4	
308008	**BRACKET, MOUNTING LOWER WISHBONE ARM ON CHASSIS**	20—53	4	
431388	**Bush,** in Bracket	20—54	4	
UFB.150/26R	**Bolt,** securing Lower Wishbone Arms to Brackets	20—55	4	
C.8667/5	**Nut,** Nyloc, on Bolts	20—56	4	
431587	**Shim,** adjusting Bracket on Chassis	20—57	As req'd	
UFS.137/11R	**Screw,** Set, securing Bracket to Chassis	20—58	8	
C.8667/3	**Nut,** Nyloc, on Bolts	20—59	8	
FW.106/T	**Washer,** Plain, under Nuts		8	
308065/05	**FRONT COIL SPRING**	20—60	2	
430581	**RUBBER SEATING FOR FRONT COIL SPRING**	20—61	4	

FRONT SUSPENSION

Part No.	Description	Plate No.	No. per Unit	Remarks
135939	**FRONT COIL SPRING SEAT ASSEMBLY**	20—62	2	
UFB.137/16R	**Bolt,** securing Seat Assembly to Lower Wishbone Arm	20—63	6	
C.8667/3	**Nut,** Nyloc, on Bolts	20—64	6	
UFS.137/8R	**Screw,** Set, securing Seat Assembly to Lower Wishbone Arm	20—65	4	
FG.106/X	**Washer,** Spring, on Setscrews	20—66	4	
431457	**RUBBER BUMP STOP ASSEMBLY**	20—67	2	
C.8667/3	**Nut,** Nyloc, securing Bump Stop to Lower Wishbone Arm	20—68	2	
307445/11	**FRONT SHOCK ABSORBER (AT.7/C)**	20—69	2	
307445/01	**Mounting Rubber** (Top)	20—70	4	
307445/02	**Washer,** outer, for Mounting Rubbers	20—71	4	
307445/03	**Washer,** centre, for Mounting Rubbers	20—72	2	
307445/04	**Distance Tube,** for Mounting Rubbers		2	
307445/05	**Nut,** at top of Shock Absorbers	20—73	2	
307445/06	**Mounting Rubber** (Bottom)	20—74	4	
430671	**BRACKET ASSEMBLY, MOUNTING FRONT SHOCK ABSORBER**	20—75	2	
430672/2	**Bracket,** for Shock Absorbers	20—76	2	
430674	**Screw,** Set, securing Shock Absorber Fulcrums	20—77	2	
430675	**Tab Washer,** on Setscrews	20—78	2	
430726	**REBOUND STOP**	20—79	2	
C.8667/3	**Nut,** Nyloc, securing Stop and Shock Absorber Bracket to Spring Seat Assembly	20—80	8	

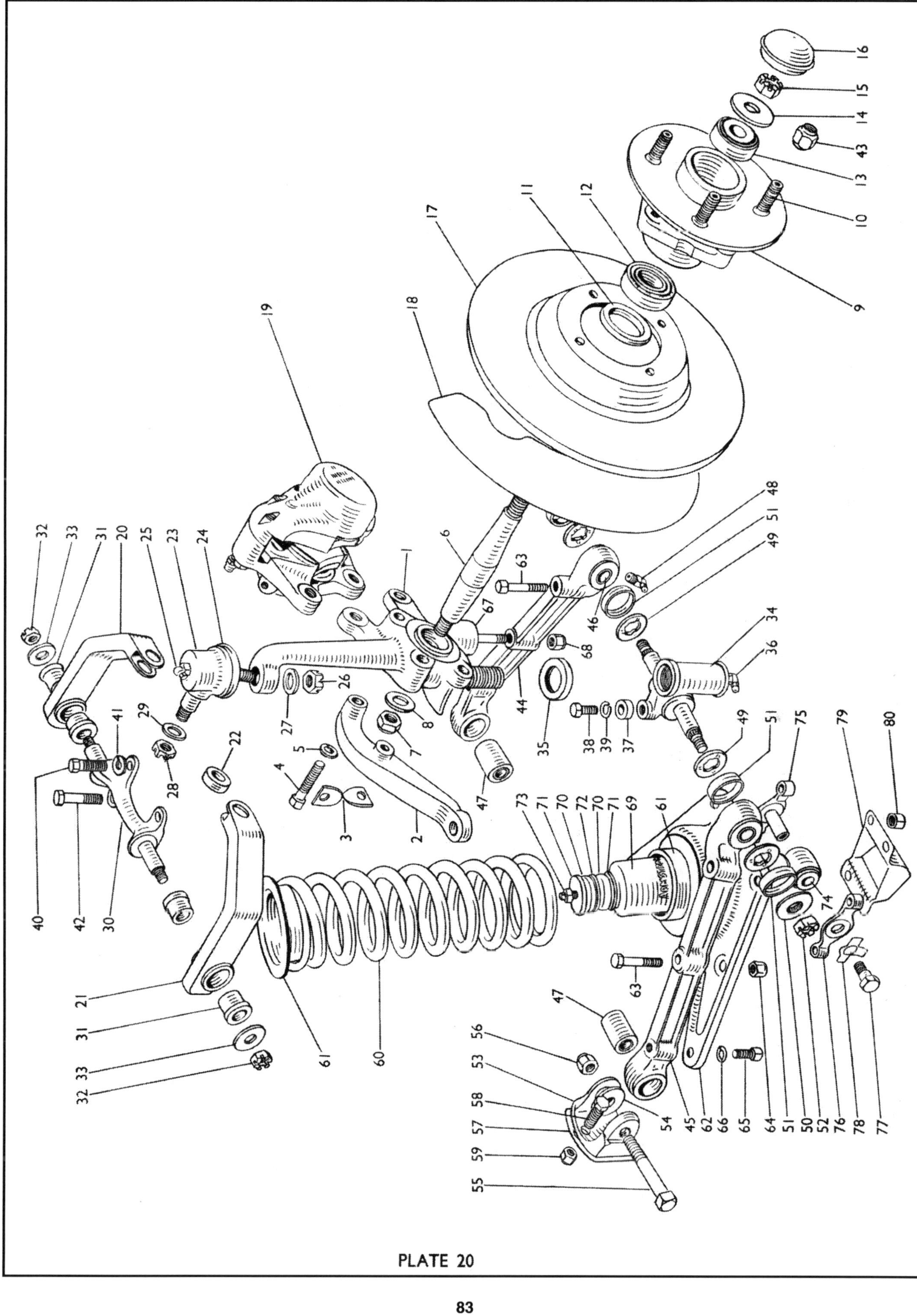

PLATE 20

ROAD WHEELS

Part No.	Description	Plate No.	No. per Unit	Remarks
136198	**DISC TYPE ROAD WHEEL (4½J. RIM x 15″ DIAMETER)**		5	
430650	**Nut,** securing Disc type Road Wheels to Hubs		16	
136201	**WIRE SPOKE ROAD WHEEL**		5	SUPPLIED TO SPECIAL ORDER ONLY
136201/01	**Long Spoke**		100	SUPPLIED TO SPECIAL ORDER ONLY
136201/02	**Short Spoke**		200	SUPPLIED TO SPECIAL ORDER ONLY
136201/03	**Nipple,** for Spokes		300	SUPPLIED TO SPECIAL ORDER ONLY
430575/01	**TYRE FOR ROAD WHEEL (5·90 x 15″)**		5	
430575/02	**TUBE FOR TYRES (5·90 x 15″)**		5	
136196	**NAVE PLATE FOR DISC TYPE ROAD WHEELS**		4	
	BALANCE WEIGHTS FOR ROAD WHEELS			
C.3985	**Balance Weight** (½ oz.)		As req'd	
C.3986	**Balance Weight** (1 oz.)		As req'd	
C.3987	**Balance Weight** (1½ ozs.)		As req'd	
C.3988	**Balance Weight** (2 ozs.)		As req'd	
C.3989	**Balance Weight** (2½ ozs.)		As req'd	
C.3990	**Balance Weight** (3 ozs.)		As req'd	
136269	**WHEEL DISC ASSEMBLY (D TYPE) FOR DISC TYPE ROAD WHEELS**		4	SUPPLIED TO SPECIAL ORDER ONLY

Part No.	Description	Plate No.	No. per Unit	Remarks
307755	**HANDBRAKE ASSEMBLY**	21—1	1	
307755/01	**Ratchet** (664/C/6024)	21—2	1	
307755/02	**Fulcrum Pin** (K.14379)	21—3	1	
307755/03	**Pawl** (446/C/4680)		1	
432558	**BRACKET FOR HANDBRAKE ASSEMBLY**		1	
UFS.137/8R	**Screw,** Set, securing Handbrake Assembly to Bracket		2	
FG.106/X	**Washer,** Spring, on Setscrews		2	
C.20817	**Screw,** Set, securing Handbrake Bracket to Chassis		2	
FG.106/X	**Washer,** Spring, on Setscrews		2	
308007/01	**HANDBRAKE CABLE ASSEMBLY**	21—4	1	
431325	**Fork End,** securing Cable to Handbrake Lever	21—5	1	
UFN.231/L	**Locknut,** securing Fork End to Cable		1	
FW.105/T	**Washer,** Packing, between Fork End and Handbrake Lever	21—6	2	
422531/17	**Clevis Pin,** securing Fork End to Handbrake Lever	21—7	1	
L.103/5U	**Split Pin,** retaining Clevis Pin		1	
FW.105/T	**Washer,** Plain, behind Split Pin	21—8	1	
308905	**ABUTMENT BRACKET AT FRONT END OF HANDBRAKE CABLE**	21—9	1	
UFS.131/5R	**Screw,** Set, securing Abutment Bracket to Chassis		2	
UFN.131/L	**Nut,** on Setscrews		2	
FG.105/X	**Washer,** Spring, under Nuts		2	
430902	**THIMBLE AT HANDBRAKE CABLE FRONT ABUTMENT**	21—10	1	
263226	**SPRING SUPPORTING HANDBRAKE CABLE AT CENTRE**	21—11	1	
428753	**Clip,** securing Spring to Cable	21—12	1	
UFS.119/3R	**Screw,** Set, securing Clip to Cable		1	
UFN.119/L	**Nut,** on Setscrew		1	
AG.102/X	**Washer,** Spring, under Nut		1	
430902	**THIMBLE, AT HANDBRAKE CABLE REAR ABUTMENT**	21—13	1	
422531/16	**CLEVIS PIN CONNECTING HAND-BRAKE CABLE TO BELL CRANK LEVER**	21—14	1	
L.102/5U	**Split Pin,** retaining Clevis Pin		1	
FW.105/T	**Washer,** Plain, behind Split Pin	21—15	1	

Part No.	Description	Plate No.	No. per Unit	Remarks
431372	**BRACKET MOUNTING BELL CRANK LEVER SWIVEL PIN ON REAR AXLE**	21—16	1	
UCS.137/6R	**Screw,** Set, securing Bracket to Rear Axle		2	
FG.106/X	**Washer,** Spring, on Setscrew		2	
431533	**BELL CRANK LEVER SWIVEL ASSEMBLY**	21—17	1	
0810/12	**Bush,** in Swivel		2	
UFN.337/L	**Slotted Nut,** securing Bell Crank Lever to Swivel Assembly	21—18	1	
L.102/7U	**Split Pin,** retaining Slotted Nut		1	
FW.106/T	**Washer,** Plain, under Slotted Nut	21—19	1	
426018	**SWIVEL PIN ASSEMBLY**	21—20	1	
UFN.350/L	**Slotted Nut,** securing Swivel Pin to Mounting Bracket	21—21	1	
L.103/8U	**Split Pin,** retaining Slotted Nut		1	
FW.106/T	**Washer,** Plain, under Slotted Nut	21—22	1	
308123	**BELL CRANK LEVER ASSEMBLY**	21—23	1	
0810/16	**Bush,** in Lever		1	
431248	**RETURN SPRING FOR BELL CRANK LEVER**	21—24	1	
431535	**BRAKE ROD, FROM BELL CRANK LEVER TO R.H. REAR BRAKE CALIPER**	21—25	1	
431534	**BRAKE ROD FROM BELL CRANK LEVER TO L.H. REAR BRAKE CALIPER**	21—26	1	
422531/03	**Clevis Pin,** securing Brake Rod to Bell Crank Lever and Calipers	21—27	4	
L.102/5U	**Split Pin,** retaining Clevis Pins		4	
FW.104/T	**Washer,** Plain, behind Split Pins at Caliper ends of Brake Rod		2	
431334	**BRAKE PEDAL ASSEMBLY**	21—28	1	R.H. DRIVE CARS ONLY
1012/11	**Bush,** in Pedal	21—29	2	

Part No.	Description	Plate No.	No. per Unit	Remarks
431397	**BRAKE PEDAL ASSEMBLY**		1	**L.H. DRIVE CARS ONLY**
1012/11	**Bush,** in Pedal		2	
427000	**RUBBER PAD ON BRAKE PEDAL**	21—30	1	
413046	**RETURN SPRING FOR BRAKE PEDAL**	21—31	1	
136118	**BRACKET MOUNTING BRAKE AND CLUTCH PEDAL**	21—32	1	
UFS.125/6R	**Screw,** Set, for adjustment of Pedal		1	
UFN.225/L	**Locknut,** securing adjusting screw		1	
90018/06	**Screw,** Set, securing Bracket to Dash		2	
UFS.125/6R	**Screw,** Set, securing Bracket to Dash		4	
C.8667/2	**Nut,** Nyloc, on setscrews		4	
51/637052	**Washer,** Plain, under Nuts		4	
430635	**FULCRUM SHAFT FOR BRAKE AND CLUTCH PEDALS**	21—33	1	
L.104/6U	**Split Pin,** retaining Fulcrum Shaft in Pedal Mounting Bracket		2	
FW.110/T	**Washer,** Plain, behind Split Pin	21—34	2	
430637	**Spacer,** between Brake and Clutch Pedals	21—35	1	
432505	**DISTANCE PIECE ON FULCRUM SHAFT**		1	**REQUIRED ONLY WHEN AUTOMATIC TRANSMISSION IS FITTED (REPLACES REDUNDANT CLUTCH PEDAL)**
308260	**BRAKE MASTER CYLINDER AND FLUID RESERVOIR ASSEMBLY (64067613)**	21—36	1	
308260/02	**Master Cylinder Assembly** (64067707)		1	
308260/01	**Brake Fluid Reservoir Assembly** (64673287)		1	
308260/05	**Adaptor,** for Fluid Reservoir (64473140)		1	
308260/06	**Gasket,** between Reservoir and Adaptor (378732)		1	
425145	**Gasket,** between Master Cylinder and Adaptor (378700)		1	
305167/02	**Nut,** securing Reservoir to Adaptor (H.4032)		1	
303996/06	**Washer,** under Nut (45-BS-5)		1	
308260/01	**BRAKE FLUID RESERVOIR ASSEMBLY (64673287)**	21—37	1	
308048/01	**Filler Cap** (378434)		1	
305167/03	**Gasket,** for Filler Cap (362920)		1	
308260/04	**Filter** (64473410)		1	
430934	**JAW, CONNECTING MASTER CYLINDER TO BRAKE PEDAL**	21—38	1	**R.H. DRIVE CARS ONLY**

BRAKING SYSTEM

Part No.	Description	Plate No.	No. per Unit	Remarks
431027	**JAW, CONNECTING MASTER CYLINDER TO BRAKE PEDAL**		1	**L.H. DRIVE CARS ONLY**
UFN.231/L	**Locknut,** securing Jaw to Master Cylinder	21—39	1	
422531/16	**Clevis Pin,** securing Jaw to Pedal	21—40	1	**R.H. DRIVE CARS ONLY**
422531/25	**Clevis Pin,** securing Jaw to Pedal		1	**L.H. DRIVE CARS ONLY**
L.103/5U	**Split Pin,** retaining Clevis Pin		1	
FW.105/T	**Washer,** Plain, behind Split Pin	21—41	1	
307973	**DISTANCE PIECE, BETWEEN BRAKE AND CLUTCH MASTER CYLINDERS AND DASH**	21—42	1	
UFB.131/24R	**Bolt,** securing Master Cylinder and Distance Piece to Dash	21—43	2	
C.8667/2	**Nut,** Nyloc, on Bolts	21—44	2	
432506	**BLANKING PLATE FOR DISTANCE PIECE**		1	**REQUIRED WHEN AUTOMATIC TRANSMISSION IS FITTED**
UFB.131/22R	**Bolt,** securing Blanking Plate and Distance Piece		2	
UFN.131/L	**Nut,** on Bolts		2	
136161	**PIPE ASSEMBLY FROM MASTER CYLINDER TO RESTRICTOR VALVE**	21—45	1	**R.H. DRIVE CARS ONLY**
136162	**PIPE ASSEMBLY FROM MASTER CYLINDER TO RESTRICTOR VALVE**		1	**L.H. DRIVE CARS ONLY**
276330/2	**Clip,** securing Brake and Clutch Pipes to Dash		3	**L.H. DRIVE CARS ONLY**
423196/43	**Screw,** Self-Tapping, securing Clips		3	
308040	**RESTRICTOR VALVE ASSEMBLY, IN 5-WAY ADAPTOR (3011060W)**	21—46	1	
431441	**Gasket,** between Restrictor Valve and 5-way Adaptor (378711W)	21—47	1	
431436	**5-WAY ADAPTOR (353811W)**	21—48	1	
431504	**Distance Piece,** between 5-way Adaptor and Chassis	21—49	1	
UFB.125/11R	**Bolt,** securing 5-way Adaptor to Chassis		1	
FG.104/X	**Washer,** Spring, on Bolt		1	
431723	**STOP LIGHT OPERATING SWITCH IN 5-WAY ADAPTOR (64067766)**		1	

Part No.	Description	Plate No.	No. per Unit	Remarks
136163	**PIPE ASSEMBLY FROM 5-WAY ADAPTOR TO L.H. FRONT BRAKE HOSE**	21—50	1	
136164	**PIPE ASSEMBLY FROM 5-WAY ADAPTOR TO R.H. FRONT BRAKE HOSE**	21—51	1	
276330/2	**Clip,** securing Pipes to Chassis		2	
425456	**Clip,** securing Pipes to Chassis		3	
423196/43	**Screw,** Self-Tapping, securing Clips		5	
308042	**FRONT BRAKE FLEXIBLE HOSE (3700822W)**	21—52	2	
426751	**Nut,** securing Flexible Hose to Front Hose Brackets (10233)		4	
425163	**Washer,** Shakeproof, under Nuts (74330)		4	
432674	**PIPE ASSEMBLY FROM FLEXIBLE HOSE TO L.H. FRONT CALIPER (64473871)**	21—53	1	
432675	**PIPE ASSEMBLY FROM FLEXIBLE HOSE TO R.H. FRONT CALIPER (64473872)**	21—54	1	
136165	**PIPE ASSEMBLY FROM 5-WAY ADAPTOR TO REAR HOSE (64473360)**	21—55	1	
425456	**Clip,** securing Pipe to Chassis		5	
423196/43	**Screw,** Self-Tapping, securing Clips		5	
305169	**REAR BRAKE FLEXIBLE HOSE FROM PIPE ASSEMBLY TO 3-WAY ADAPTOR (3700625W)**	21—56	1	
426751	**Nut,** securing Hose to Bracket	21—57	1	
426163	**Washer,** Shakeproof, under Nut	21—58	1	
431441	**Gasket,** between Hose and 3-way Adaptor (378711)	21—59	1	
431442	**3-WAY ADAPTOR (353231W)**	21—60	1	
UFB.125/10R	**Bolt,** securing 3-way Adaptor to Swivel Bracket on Rear Axle		1	
C.8667/2	**Nut,** Nyloc, on Bolt		1	

Part No.	Description	Plate No.	No. per Unit	Remarks
136166	**PIPE ASSEMBLY, FROM 3-WAY ADAPTOR TO L.H. REAR CALIPER (64473380)**	21—61	1	
276300/3	**Clip,** securing Pipe to Rear Axle Banjo		1	
302454/36	**Clip,** securing Pipe to Rear Axle		1	
136167	**PIPE ASSEMBLY, FROM 3-WAY ADAPTOR TO R.H. REAR CALIPER (64473379)**	21—62	1	
302454/31	**Clip,** securing Pipe to Rear Axle		1	
308260/02	**BRAKE MASTER CYLINDER ASSEMBLY (64067707)**		1	
308048/04	**Spring,** returning Plunger (64673425)	22—63	1	
308048/02	**Push Rod,** (350527)	22—64	1	
7352	**Washer,** returning Push Rod (378230)	22—65	1	
308260/07	**BRAKE MASTER CYLINDER REPAIR KIT (SP.2095)**	22—66	1	
	Comprising:—			
	Valve Seal			
	Plunger Seal			
	Gasket, for Reservoir Filler Cap			
	Dust Cover			
	Circlip			
	Washer, for Valve Spring			
	Tube of Grease			
136157	**L.H. FRONT BRAKE CALIPER ASSEMBLY (64032500 C.V.)**		1	
136158	**R.H. FRONT BRAKE CALIPER ASSEMBLY (64032501 C.V.)**		1	
136157/01	**Pad Assembly** (64325013 CV)	22—67	4	
136157/02	**Retaining Pin** (64325022)	22—68	4	
136157/03	**Clip,** for Retaining Pin (64325009)	22—69	4	
136157/04	**Bleed Screw,** (377687W)	22—70	2	
134617/05	**Dust Cap,** on Bleed Screw (363400)	22—71	2	
136157/05	**FRONT CALIPER REPAIR KIT (SP.2501)**	22—72	1	
	Comprising:—			
	Cylinder Sealing Ring			
	Dust Cover			

Part No.	Description	Plate No.	No. per Unit	Remarks
10130	**FRICTION PAD KIT FOR FRONT CALIPER (64325424 D.T.)**		1	
136204	**L.H. REAR BRAKE CALIPER ASSEMBLY (64032522. XA)**		1	
136203	**R.H. REAR BRAKE CALIPER ASSEMBLY (64032523. XA)**		1	
136203/01	**Pad Assembly** (64325056 CV)	22—73	4	
136157/02	**Retaining Pin** (64325022)	22—74	4	
136157/03	**Clip,** for Retaining Pin (64325009)	22—75	4	
136157/04	**Bleed Screw** (377687W)	22—76	2	
134617/05	**Dust Cap,** on Bleed Screw (363400)	22—77	2	
136204/01	**Handbrake Pad Assembly** L.H. (64325568 EE)	22—78	2	
136203/02	**Handbrake Pad Assembly,** R.H. (64325569 EE)	22—78A	2	
136203/03	**Clamping Lever** (Short) (64325196)	22—79	2	
136203/04	**Clamping Lever** (Long) (64325195)	22—80	2	
136203/05	**Operating Lever Assembly** (64325410)	22—81	2	
136203/07	**Pivot Pin,** for Clamping Lever (64325194)	22—82	4	
136203/08	**Pivot Pin,** for Operating Lever (64325211)	22—83	2	
FW.105/T	**Washer,** on Operating Lever Pivot Pin (45-BS-2)	22—84	4	
136203/09	**Tie Rod,** for Handbrake Lever (64325197)	22—85	2	
136203/10	**Spring,** on Tie Rod (64325198)	22—86	2	
FW.105/T	**Washer,** on Tie Rod (45-BS-2)	22—87	2	
136203/11	**Adjuster Nut,** for Handbrake (64325201)	22—88	2	
136203/12	**Locking Spring,** for Adjuster Nut (64325200)	22—89	2	
136203/06	**REAR CALIPER REPAIR KIT (SP.2505)**		1	
	Comprising:—			
	Cylinder Sealing Ring			
	Dust Cover			
136203/13	**FRICTION PAD KIT FOR HANDBRAKE (64932004 EE)**		1	
10131	**FRICTION PAD KIT FOR REAR CALIPER (64325422 D.T.)**		1	

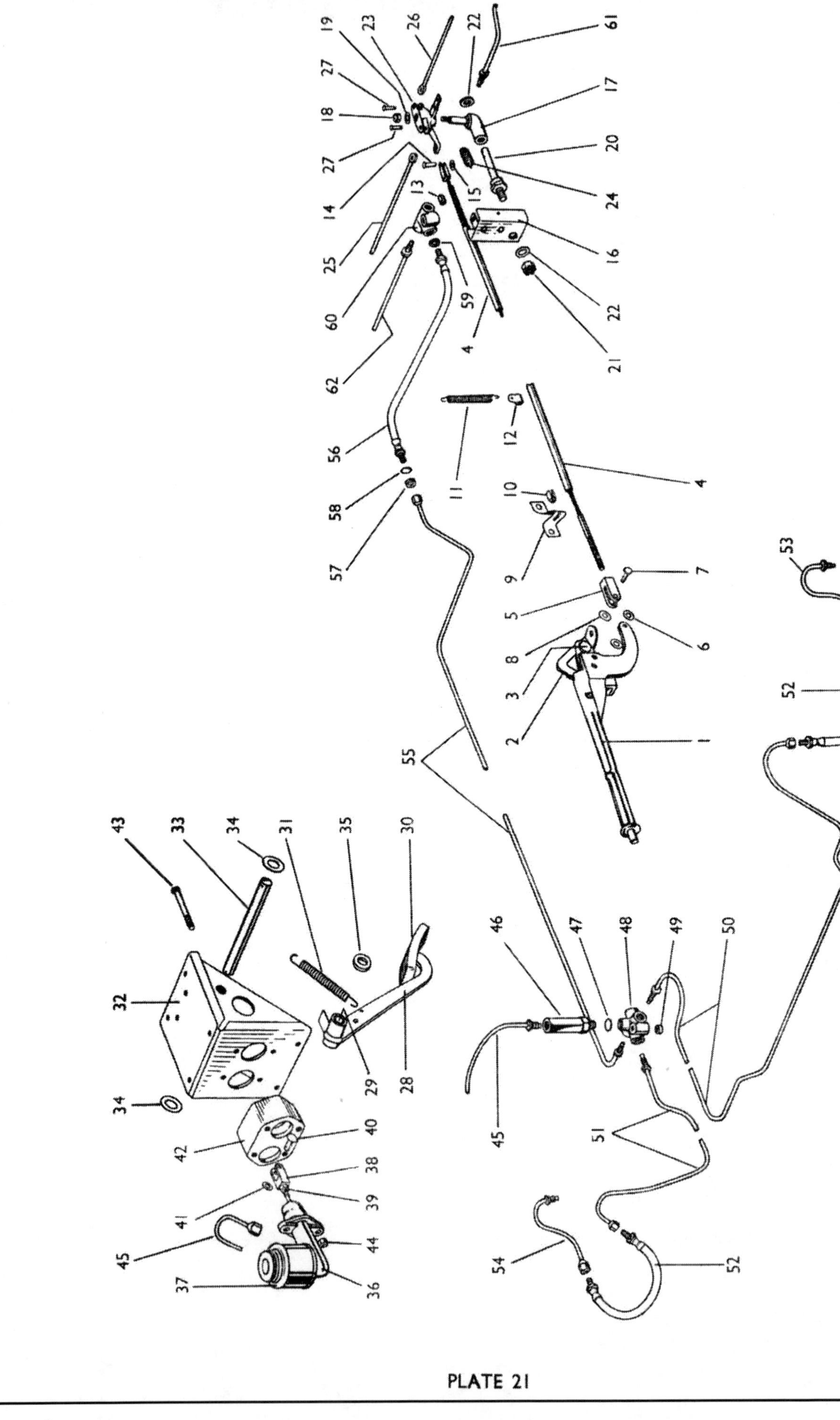
19
23
26
22
61
27
18
17
27
20
14
13
15
24
25
16
60
59
22
62
4
21
56
11
12
58
10
4
57
9
7
5
6
8
3
2
1
53
52
55
43
33
34
31
35
30
46
47
48
49
50
32
29
28
45
34
42
40
38
39
41
51
44
45
37
36
54
52

PLATE 21

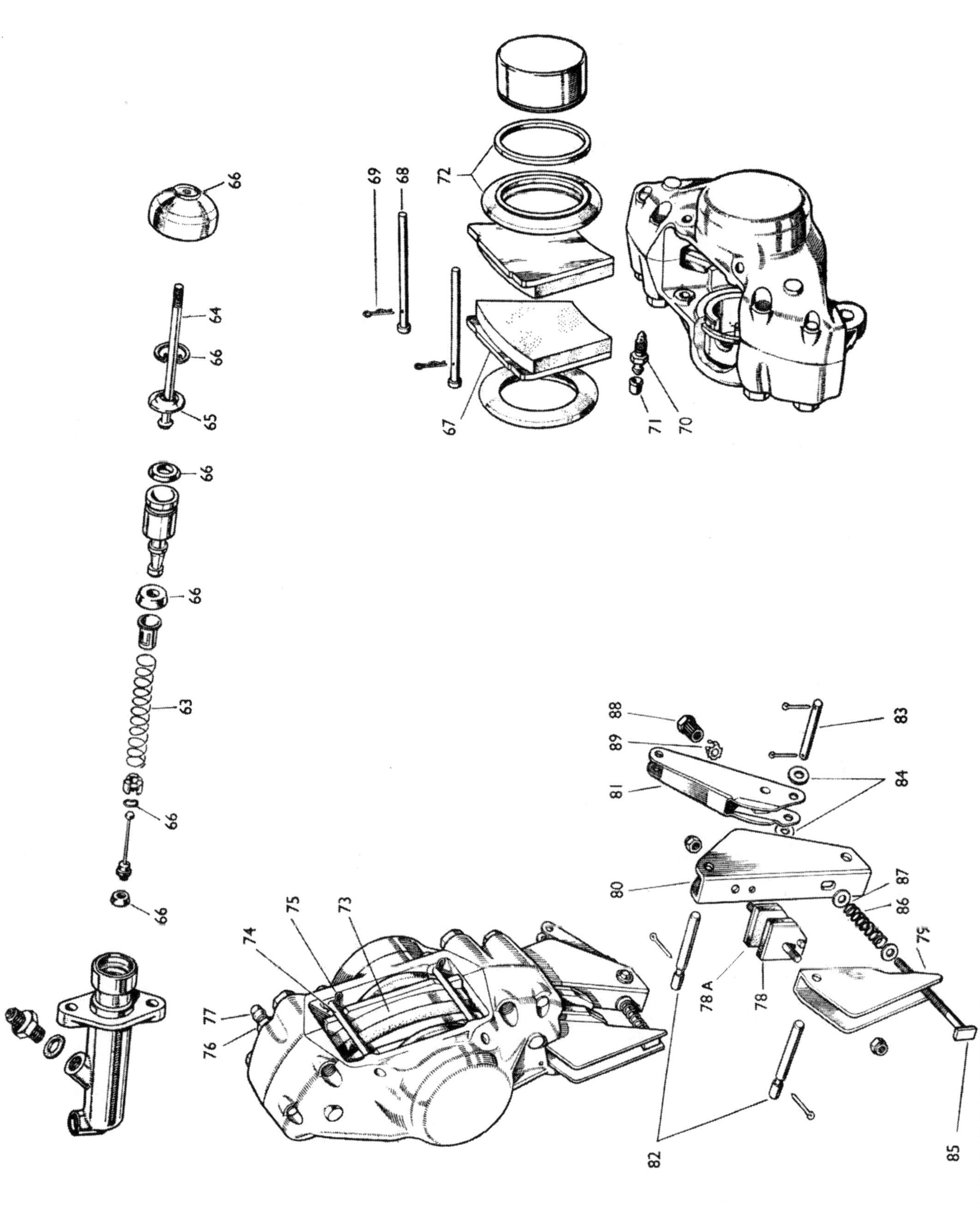

63
64
65
66
67
68
69
70
71
72
73
74
75
76
77
78
78 A
79
80
81
82
83
84
85
86
87
88
89

PLATE 22

STEERING UNIT AND CONTROLS

Part No.	Description	Plate No.	No. per Unit	Remarks
135932	**STEERING COLUMN CONTROL ASSEMBLY (33549)**	23—1	1	
135932/01	**Top Cover Assembly** (54300363)	23—2	1	
135932/02	**Horn Push** (54300354)		1	
135932/03	**Retaining Ring** for Horn Push (324366)		1	
133055/03	**Spring,** for Horn Push (322093)		1	
2842	**Contact,** for Horn Push (321824)		1	
135246/03	**Lever,** for Direction Indicator Switch (324992)	23—3	1	
2866	**Base,** for Direction Indicator Switch (323438)		1	
2843	**Plunger,** for Rotor (322150)		1	
2877	**Cam Assembly** (324527)		1	
135932/05	**Stator Tube and Plate** (54300006)	23—4	1	
426383	**Spring,** on Stator Tube (324072/D)	23—5	3	
135932/06	**Cable** (54940043)		1	
135990	**STEERING WHEEL**	23—6	1	
427234	**Nut,** securing Steering Wheel	23—7	1	
427235	**Washer,** Plain, under Nut	23—8	1	
405066	**Striker Plate,** for Trafficator Switch	23—9	1	
136057	**STEERING BOX ASSEMBLY COMPLETE WITH OUTER COLUMN (Z.1385)**		1	**R.H. DRIVE CARS ONLY**
136057/01	**Steering Box** (complete with Trunnion Bush and Oil Seal) (37871)	23—10	1	
135217/04	**Bush** (Felt) in top of Outer Column (16157)	23—12	1	
136058/02	**Oil Plug** (Rubber) in Outer Column (15776)	23—13	1	
136057/02	**Inner Column and Worm,** (complete with Oil Seals) (37760)	23—14	1	
135217/06	**Inner Race,** for Inner Column Worm (7712)	23—16	2	
135217/07	**Outer Race,** for Inner Column Worm (8527)	23—17	2	
133502/20	**End Plate,** for Steering Box (17312)	23—18	1	
133502/23	**Gasket,** under End Plate (9420)	23—19	1	
133502/24	**Shim** (·0024" thick) under End Plate, for adjustment of Worm End Float (8224)	23—20	As req'd	
133502/25	**Shim** (·005" thick) under End Plate, for adjustment of Worm End Float (9418)	23—20	As req'd	
133502/26	**Shim** (·010" thick) under End Plate, for adjustment of Worm End Float (9419)	23—20	As req'd	
UFB.131/5R	**Bolt,** securing End Plate (6448U)	23—21	4	
FG.105/X	**Washer,** Spring, on Bolts (6385)	23—22	4	
136058/05	**Olive,** on Stator Tube	23—23	1	
133502/21	**Union Nut,** securing Stator Tube at End Plate	23—24	1	
136058/06	**Rocker Shaft Assembly** (27278U)	23—25	1	
136057/03	**Cover Plate,** at bottom of Steering Box (7787)	23—26	1	
136058/08	**Screw,** in Cover Plate, for adjustment of Rocker Shaft (16983)	23—27	1	
135217/15	**Locknut,** on adjusting screw (12756)	23—28	1	
136058/11	**Gasket,** under Cover Plate (10732)	23—31	1	
UFB.131/5R	**Bolt,** securing Cover Plate (6448U)	23—32	3	
	Washer, Spring, on Bolts	23—33	3	
307775	**Drop Arm,** on Rocker shaft (18423)	23—34	1	
136058/12	**Nut,** securing Drop Arm (16146 U)	23—35	1	
136058/11	**Lockwasher,** securing Nut, (14506)	23—36	1	
136058	**STEERING BOX ASSEMBLY, COMPLETE WITH OUTER COLUMN (Z.1384)**		1	**L.H. DRIVE CARS ONLY**
136058/01	**Steering Box** (complete with Trunnion Bush and Oil Seal) (37875)		1	

Part No.	Description	Plate No.	No. per Unit	Remarks
	STEERING BOX ASSEMBLY (Cont.)			
136058/03	**Inner Column and Worm** (complete with Oil Seals) (37761)		1	
308193	**Drop Arm,** on Rockershaft		1	
	All other items are as for Steering Box Assembly Part No. 136057			
136057/01	**STEERING BOX (COMPLETE WITH TRUNNION BUSH AND OIL SEAL) (37871)**	23—10	1	**R.H. DRIVE CARS ONLY**
135217/02	**Trunnion Bush** (8129)		1	
135217/03	**Oil Seal** (15535)	23—11	1	
136058/01	**STEERING BOX (COMPLETE WITH TRUNNION BUSH AND OIL SEAL) (37875)**		1	**L.H. DRIVE CARS ONLY**
135217/02	**Trunnion Bush** (8129)		1	
135217/03	**Oil Seal** (15535)		1	
136057/02	**INNER COLUMN AND WORM (COMPLETE WITH OIL SEALS) (37760)**	23—14	1	**R.H. DRIVE CARS ONLY**
136058/04	**Oil Seal** (Rubber) (8648)	23—15	2	
136058/03	**INNER COLUMN AND WORM (COMPLETE WITH OIL SEALS) (37761)**		1	**L.H. DRIVE CARS ONLY**
136058/04	**Oil Seal** (Rubber) (8648)		2	
432204	**Oil Filler Nipple,** in Steering Box (Tec. N.6625)		1	
432202	**Tag Cover,** for Nipple		1	
431154	**BRACKET ASSEMBLY, MOUNTING IDLER LEVER**	23—37	1	
431606	**Grease Nipple,** in Bracket	23—38	1	
UFB.137/52R	**Bolt,** securing Bracket to Chassis		2	
C.8667/3	**Nut,** Nyloc, on Bolts		2	
431153	**IDLER LEVER ASSEMBLY**	23—39	1	
431605	**Grease Seal,** between Lever and Bracket	23—40	1	

Part No.	Description	Plate No.	No. per Unit	Remarks
307782	**CENTRE TIE ROD ASSEMBLY**	23—41	1	
136054	**Centre Tie Rod**	23—42	1	
431400	**Bush Assembly** (Silentbloc)	23—43	2	
FW.106/T	**Washer,** Plain, under Nuts	23—44	2	
308198/01	**Nut,** Nyloc, securing Tie Rod to Idler Lever and Drop Arm	23—45	2	
308022	**R.H. OUTER TIE ROD ASSEMBLY**	23—46	1	
431600	**Outer Tie Rod R.H.**	23—47	1	
431601	**Locknut** (right-hand thread) securing Inner Ball Joint Assembly	23—48	1	
431602	**Locknut** (left-hand thread) securing Outer Ball Joint Assembly	23—49	1	
308199	**Inner Ball Joint Assembly** (right-hand thread)	23—50	1	
308201	**Outer Ball Joint Assembly** (left-hand thread)	23—51	1	
308199	**INNER BALL JOINT ASSEMBLY FOR R.H. OUTER TIE ROD**		1	
431592	**Gaiter,** for Ball Joint	23—52	1	
431603	**Retainer,** inside Gaiter		1	
C.9048	**Grease Nipple,** for Ball Joint	23—53	1	
308198/01	**Nut,** Nyloc, on Ball Joint	23—54	1	
FW.106/T	**Washer,** Plain, under Nut	23—55	1	
308201	**OUTER BALL JOINT ASSEMBLY FOR R.H. OUTER TIE ROD**		1	
	Individual items are as for Inner Ball Joint Assembly 308199			
308023	**L.H. OUTER TIE ROD ASSEMBLY**	23—56	1	
431600	**Outer Tie Rod L.H.**	23—57	1	
431601	**Locknut,** (right-hand thread) securing Inner Ball Joint Assembly	23—58	1	
431602	**Locknut,** (left-hand thread) securing Outer Ball Joint Assembly	23—59	1	
308198	**Inner Ball Joint Assembly** (right-hand thread)	23—60	1	
308200	**Outer Ball Joint Assembly** (left-hand thread)	23—61	1	
308198	**INNER BALL JOINT ASSEMBLY FOR L.H. OUTER TIE ROD**		1	
431592	**Gaiter,** for Ball Joint	23—62	1	
431603	**Retainer,** inside Gaiter		1	
C.9048	**Grease Nipple,** for Ball Joint	23—63	1	
308198/01	**Nut,** Nyloc, on Ball Joint		1	
FW.106/T	**Washer,** Plain, under Nut		1	
308200	**OUTER BALL JOINT ASSEMBLY FOR L.H. OUTER TIE ROD**		1	
	Individual items are as for Inner Ball Joint Assembly 308198			

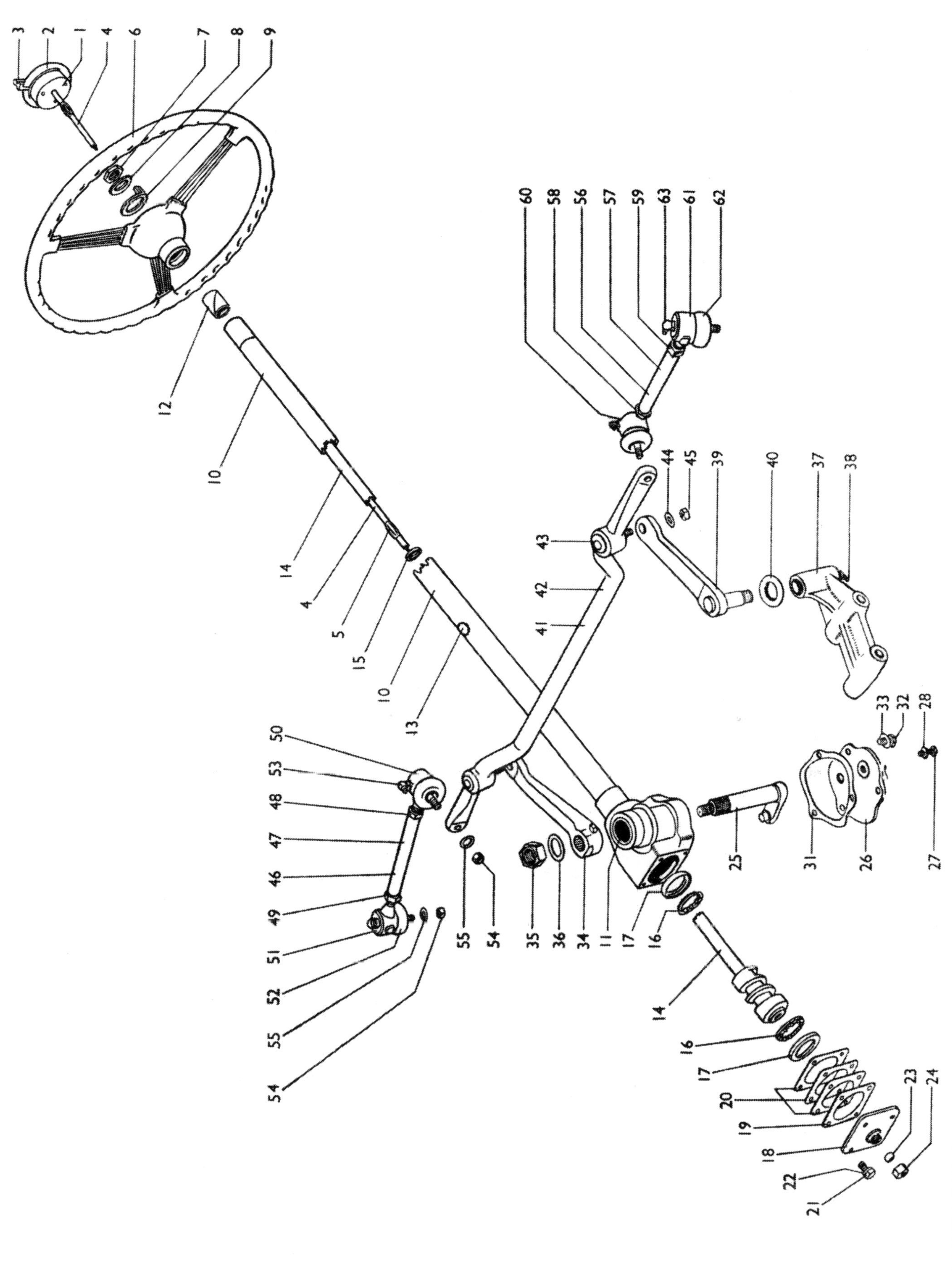
3 2 1 4 6 7 8 9
60 58 56 57 59 63 61 62
12 10 14 4 5 15 10 13
43 42 41 44 45 39 40 37 38
33 32 28
50 53 48 47 46 49 51 52 55 54
55 54 35 36 34 11 17 16
25 31 26 27
14 16 17 20 19 18 22 21 23 24

PLATE 23

COOLING SYSTEM

Part No.	Description	Plate No.	No. per Unit	Remarks
135931	**RADIATOR BLOCK ASSEMBLY**	24—1	1	
UFB.131/6R	**Bolt,** securing Radiator Block		4	
FG.105/X	**Washer,** Spring, on Bolts		4	
FW.105/T	**Washer,** Plain, under Spring Washers		4	
422505/11	**DRAIN TAP, FOR RADIATOR BLOCK**	24—2	1	
424775	**Copper Washer,** on Drain Tap	24—3	1	
307980	**HEADER TANK ASSEMBLY**	24—4	1	
431362	**HOSE, FROM HEADER TANK TO WATER OUTLET PIPE**	24—5	1	
431363	**HOSE, FROM RADIATOR BLOCK TO HEADER TANK**	24—6	1	
423176/23	**Clip,** securing Hoses	24—7	4	
431361	**OVERFLOW PIPE ON HEADER TANK**	24—8	1	
428753	**Clip,** securing Overflow Pipe to Valance		1	
425286/17	**Screw,** securing Clip		1	
308272	**FILLER CAP ON HEADER TANK (1571515)**	24—9	1	
307983	**HOSE, FROM WATER PUMP TO BOTTOM OF RADIATOR BLOCK**	24—10	1	
423176/23	**Clip,** securing Hose to Water Pump and Radiator Block	24—11	2	

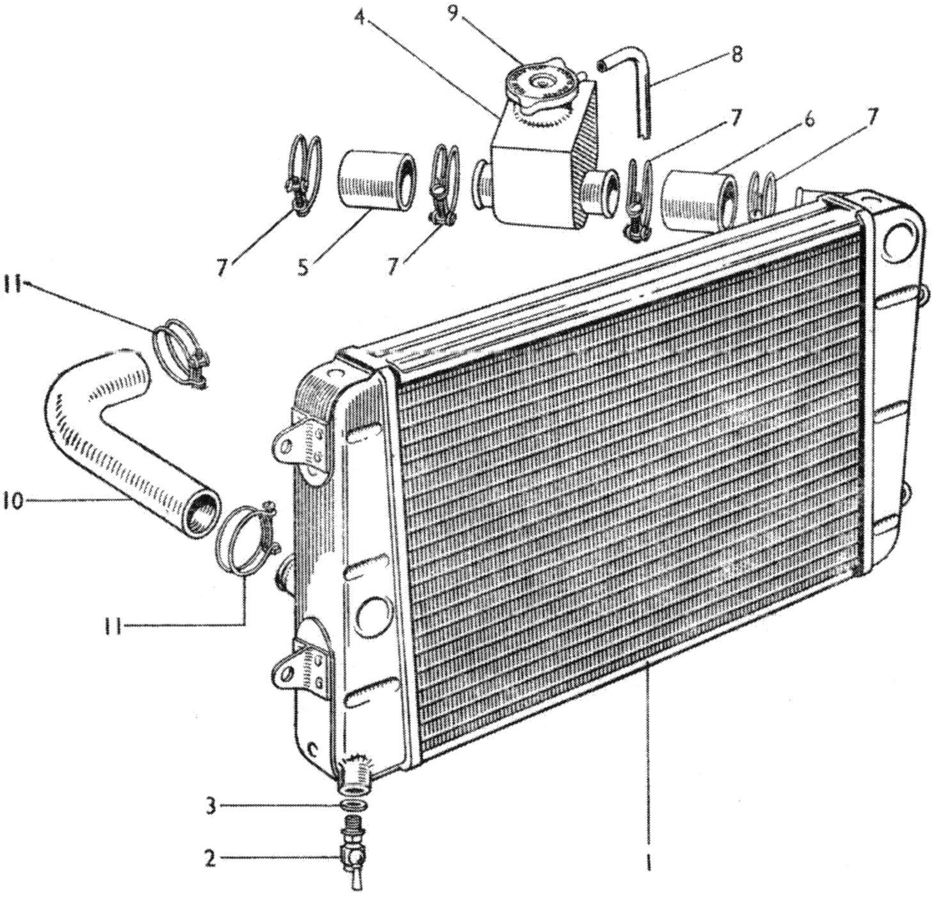

PLATE 24

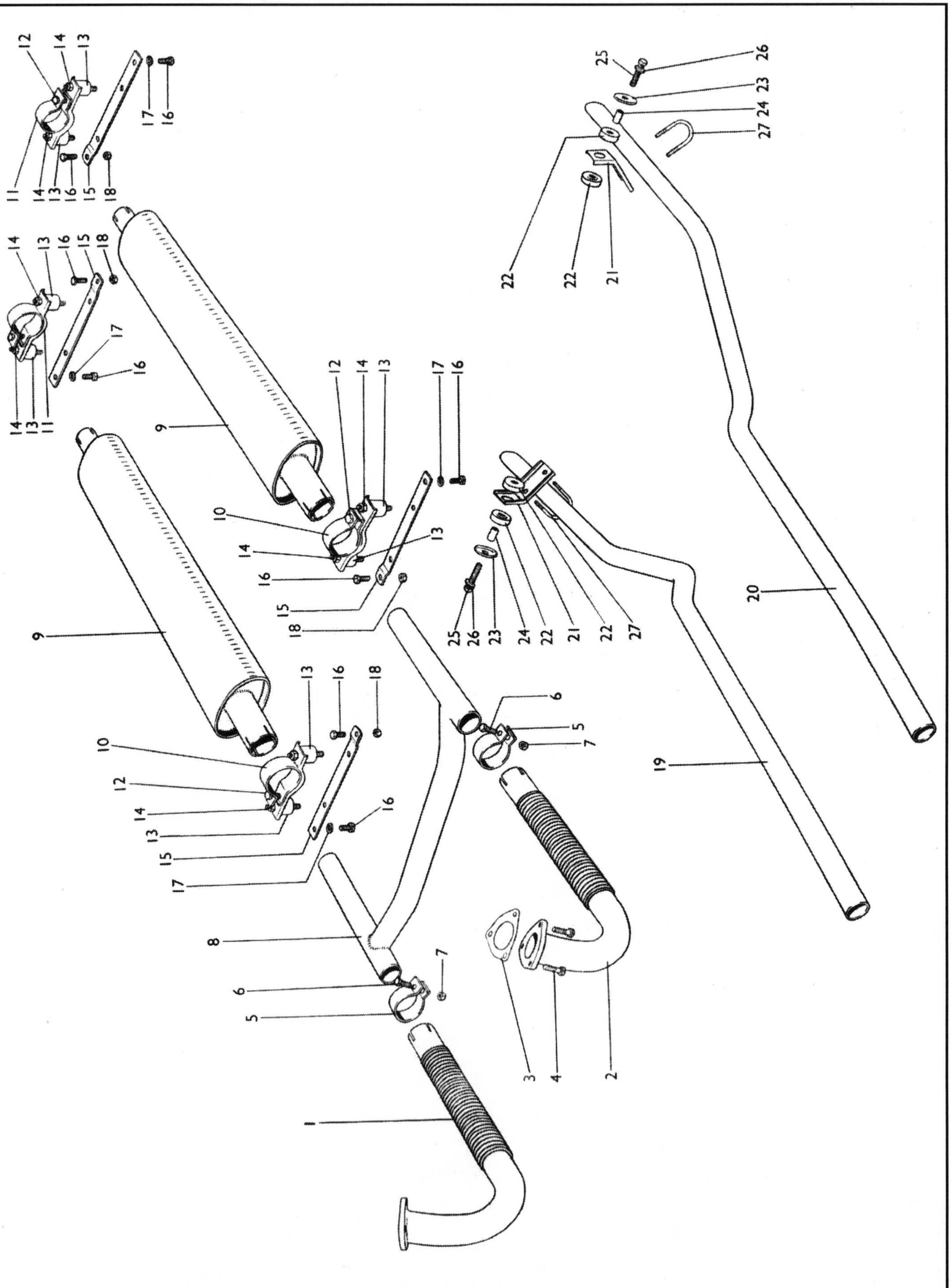

PLATE 25

Part No.	Description	Plate No.	No. per Unit	Remarks
136169	**R.H. EXHAUST DOWN PIPE ASSEMBLY**	25—1	1	
431447	**Flexible Pipe,** on Down Pipe		1	
136170	**L.H. EXHAUST DOWN PIPE ASSEMBLY**	25—2	1	
431447	**Flexible Pipe,** on Down Pipe		1	
422468	**Gasket,** between Down Pipes and Exhaust Manifolds	25—3	2	
UFB.131/10R	**Bolt,** securing Down Pipes to Exhaust Manifolds	25—4	6	
422512/06	**Nut,** Self-Locking, on Bolts		6	
431094	**Clip,** securing Balance Pipe to Down Pipes	25—5	2	
UFB.131/6R	**Bolt,** securing Clip	25—6	2	
422512/06	**Nut,** Self-Locking, on Bolts	25—7	2	
308053	**EXHAUST BALANCE PIPE ASSEMBLY**	25—8	1	NOT REQUIRED WHEN AUTOMATIC TRANSMISSION IS FITTED
308320	**EXHAUST BALANCE PIPE ASSEMBLY**		1	REQUIRED WHEN AUTOMATIC TRANSMISSION IS FITTED
308064	**EXHAUST SILENCER ASSEMBLY**	25—9	2	
431346	**Clip Assembly,** securing Silencer to Balance Pipe	25—10	2	
431347	**Clip Assembly,** securing Silencer to Tail Pipe Assembly	25—11	2	
UFB.131/8R	**Bolt,** securing Clips	25—12	2	
422512/06	**Nut,** Self-Locking on Bolts		2	
431100	**Packing Piece**		2	FOR USE WITH PT. NO. 431347
431339	**Support Rubber Assembly,** securing Clip Assemblies to Support Plates	25—13	8	
UFN.131/L	**Nut,** securing Support Rubbers	25—14	16	
906083	**Palnut,** behind Nuts		16	
307884	**Support Plate,** for Clip Assemblies	25—15	4	
UFS.131/6R	**Screw,** Set, securing Support Plate to Chassis	25—16	8	
FG.105/X	**Washer,** Spring, under Nuts	25—17	4	
422512/05	**Nut,** on Setscrews at outer ends of Support Plates	25—18	4	
FW.105/T	**Washer,** Plain, under Spring Washers		4	
136174	**R.H. EXHAUST TAIL PIPE ASSEMBLY**	25—19	1	
136175	**L.H. EXHAUST TAIL PIPE ASSEMBLY**	25—20	1	
431461	**EXHAUST TAIL PIPE HANGER BRACKET**	25—21	2	
425188	**Rubber Bush,** on Bolts at each side of Brackets	25—22	4	
425187	**Retaining Washer,** for Outer Bushes	25—23	2	
425189	**Distance Tube,** through Bushes and Brackets	25—24	2	
UFB.137/14R	**Bolt,** securing Brackets to Chassis	25—25	2	
FG.106/X	**Washer,** Spring, on Bolts	25—26	2	
423656	**'U' Bolt,** securing Tail Pipe Assemblies to Hanger Brackets	25—27	2	
422512/05	**Nut,** on 'U' Bolts		4	

EXHAUST SYSTEM

Part No.	Description	Plate No.	No. per Unit	Remarks
81/634364	**HEAT SHIELD (ASBESTOS) AT FRONT OF EXHAUST SILENCERS**		2	
81/644365	**COVER PLATE, FOR HEAT SHIELDS**		2	
81/634363	**HEAT SHIELD (ASBESTOS) AT REAR OF EXHAUST SILENCERS**		2	
81/644366	**COVER PLATE FOR HEAT SHIELDS**		2	
423195/38	**Screw,** Set, securing Heat Shields and Cover Plates		16	
423199/2	**Washer,** Plain, on Setscrews		16	

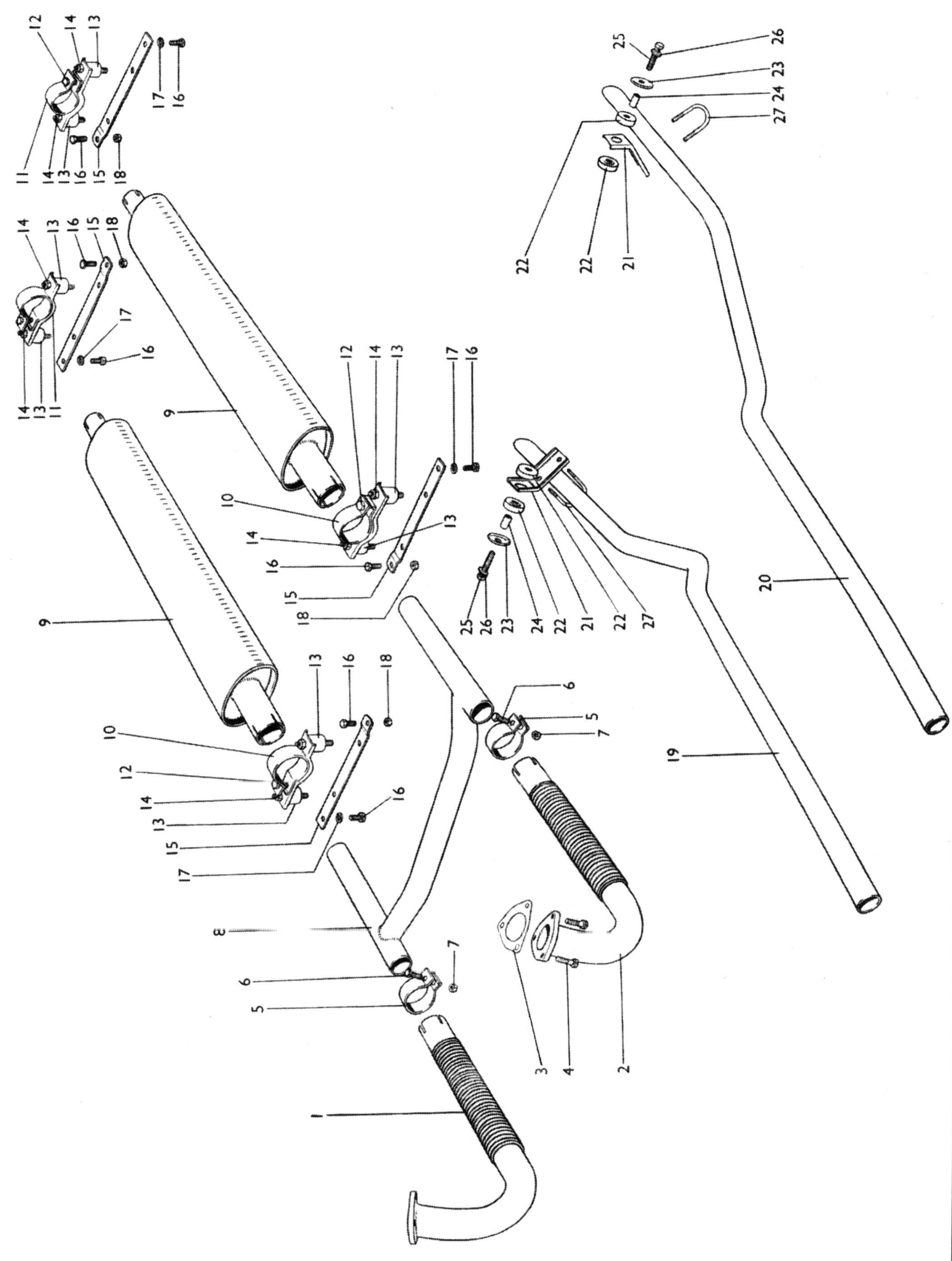

PLATE 25

FUEL SYSTEM

Part No.	Description	Plate No.	No. per Unit	Remarks
136116	**PETROL TANK ASSEMBLY**	26—1	1	
426867	**Drain Plug,** in Petrol Tank	26—2	1	
432530/04	**Copper, Washer,** on Drain Plug	26—3	1	
307987	**PETROL TANK FEED PIPE ASSEMBLY**	26—4	1	
431352	**Gasket,** under Flange of Feed Pipe Assembly	26—5	1	
426934/25	**Screw,** securing Feed Pipe Assembly to Petrol Tank		2	
431497	**Washer,** Fibre, on Screws		2	
425599	**FILLER CAP**	26—6	1	
431357	**FILLER NECK FOR PETROL TANK**	26—7	1	
425610	**Sealing Washer,** for Filler Neck		1	
419522	**Grommet,** in Body for Filler Neck	26—8	1	
431359	**HOSE CONNECTING FILLER NECK TO PETROL TANK**	26—10	1	
423176/30	**Clip,** securing Hose to Filler Neck and Petrol Tank		2	
81/634105	**STRAP (TOP) SECURING PETROL TANK TO BODY**		2	
81/632272	**STRAP (REAR) SECURING PETROL TANK TO BODY**		2	
UFS.131/7R	**Screw,** Set, securing Top Strap to Rear Strap		2	
C.8667/2	**Nut,** on Setscrews		2	
11405/04	**Bolt,** securing Petrol Tank Straps to Body		4	
C.725	**Washer,** Shakeproof, on Bolts		4	
307994	**PETROL TANK ELECTRIC ELEMENT UNIT (FT.5301/05)**		1	
307994/01	**Gasket,** under Element Unit		1	
426934/22	**Screw,** securing Element Unit to Petrol Tank		6	
431467	**HOSE ASSEMBLY FROM PETROL TANK FEED PIPE TO PETROL PUMP**	26—11	1	
302454/3	**Clip,** securing Hose to Feed Pipe		1	
432278	**Grommet,** in Body, for Hose	26—12	1	

Part No.	Description	Plate No.	No. per Unit	Remarks
C.9688/1	**PETROL PUMP (AUA 52)**	26—13	1	**1st CONDITION**
C.18336	**PETROL PUMP (AUA 152)**	26—13	1	**2nd CONDITION**
C.20576	**PETROL PUMP (AUB 152)**	26—13	1	**3rd CONDITION**
308165	**BRACKET MOUNTING PETROL PUMP**	26—14	1	
UFS.131/6R	**Screw,** Set, securing Bracket to Body		2	
FG.105/X	**Washer,** Spring, on Setscrews		2	
422284	**Grommet,** in Bracket	26—15	2	
431480	**Distance Tube,** in Grommets	26—16	2	
431481	**Washer,** Plain, at each side of Grommets	26—17	4	
900155	**Bolt,** securing Petrol Pump to Bracket	26—18	2	
FG.105/X	**Washer,** Spring, on Bolts	26—19	2	
431466	**FLEXIBLE PIPE ASSEMBLY FROM PETROL PUMP TO REAR PIPE ASSEMBLY**	26—20	1	
432278	**Grommet,** in Body for Flexible Pipe	26—21	2	
302454/3	**Clip,** securing Flexible Pipe to Rear Petrol Pipe Assembly		1	
308068	**PETROL PIPE ASSEMBLY FROM REAR FLEXIBLE PIPE TO PETROL FILTER**	26—22	1	**SUPERSEDED BY LATER CONDITION SEE BELOW**
276302/8	**Clip,** securing Pipe to Chassis	26—23	7	
423196/43	**Screw,** securing Clip		7	
414273	**Clip,** securing Flexible Pipe to Shield	26—24	1	
426934/35	**Screw,** securing Clip		1	
UFN.119/L	**Nut,** on Screw		1	
AG.102/X	**Washer,** Spring, under Nut		1	
405402	**Grommet,** in Valance for Flexible Pipe	26—25	1	
414273	**Clip,** securing Flexible Pipe to Valance	26—26	1	
425286/17	**Screw,** securing Clip		1	
422787	**Union,** connecting Petrol Pipe to Petrol Filter	26—27	1	
309216	**REAR PETROL PIPE ASSEMBLY**		1	**SUPERSEDES EARLIER CONDITION**
276302/8	**Clip,** securing Pipe to Chassis		7	
423196/43	**Screw,** securing Clips		7	
136686	**FRONT PETROL PIPE ASSEMBLY**		1	
432278	**Grommet,** in Valance for Pipe		1	
276302/8	**Clip,** securing Pipe to Heat Shield		1	
426934/35	**Screw,** securing Clip		1	
426935/4	**Nut,** on Screw		1	
FG.103/X	**Washer,** Spring, under Nut		1	

FUEL SYSTEM

Part No.	Description	Plate No.	No. per Unit	Remarks
307988	**PETROL FILTER ASSEMBLY** (7950290)	26—28	1	
307988/01	**Body,** for Filter (7950289)	26—29	1	
133007/13	**Gasket,** between Glass Bowl and Filter Body (5593321)	26—30	1	
133007/14	**Gauze,** for Filter (1524849)	26—31	1	
307988/02	**Sealing Ring** for 'ON/OFF' Valve (1524845)		1	
133007/12	**Glass Bowl,** (1523620)	26—32	1	
133007/11	**Retainer Assembly,** for Glass Bowl (5592050)	26—33	1	
90018/06	**Screw,** Set, securing Filter Assembly to Valance		2	
432974/02	**PETROL PIPE FROM FILTER TO CARBURETTERS**	26—34	1	1st CONDITION
431476	**Union,** securing Petrol Pipe to Filter		1	1st CONDITION
C.20317/02	**PETROL PIPE ASSEMBLY FROM FILTER TO CARBURETTERS**		1	2nd CONDITION
C.20295	**Olive,** on Pipe		2	2nd CONDITION
C.20294	**Union Nut,** securing Pipe to Filter and Carburetters		2	2nd CONDITION

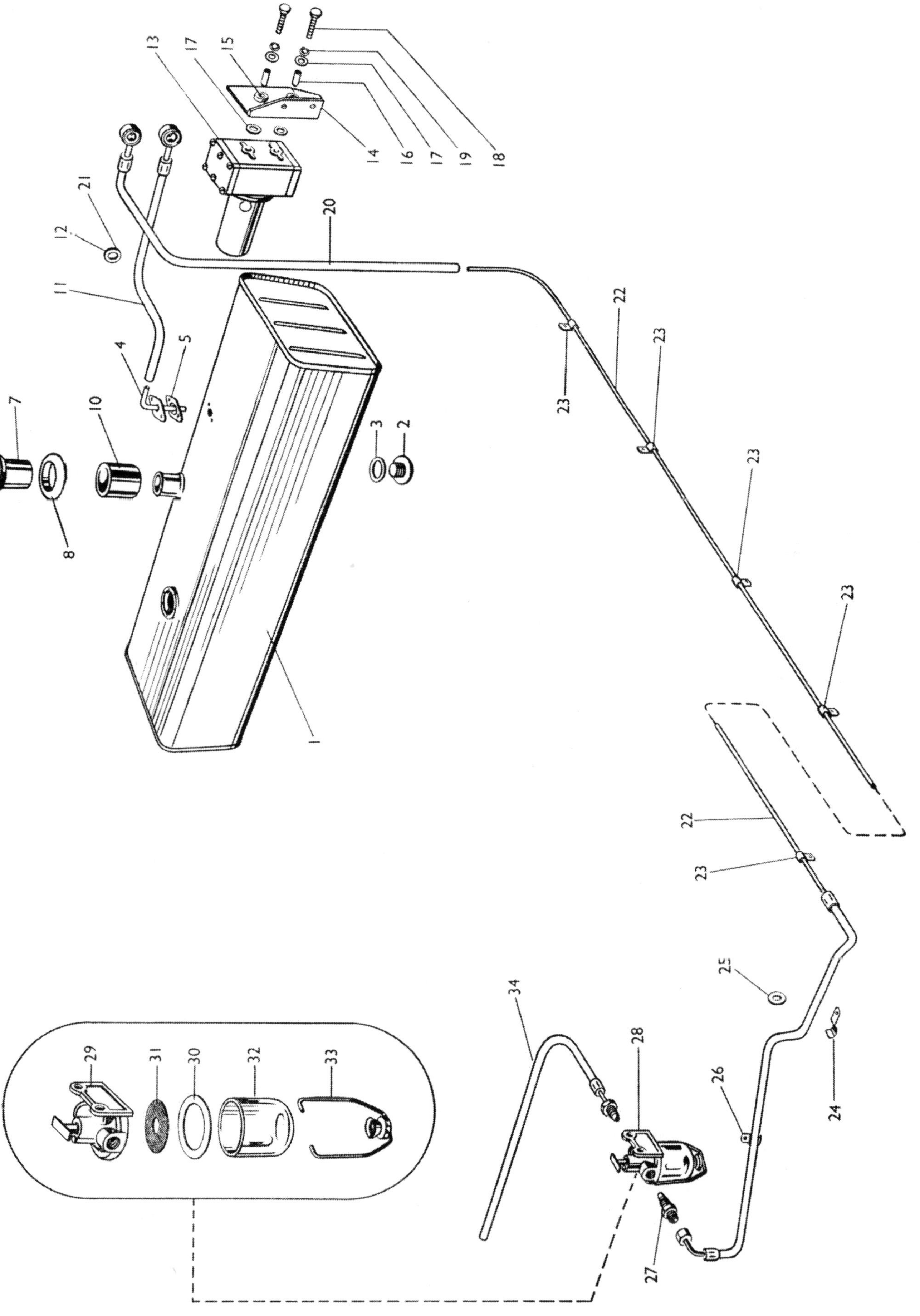
13
17
15
14
16
17
19
18
21
12
20
11
4
5
6
7
10
8
3
2
1
22
23
23
23
23
23
22
23
25
24
26
27
28
34
29
31
30
32
33

PLATE 26

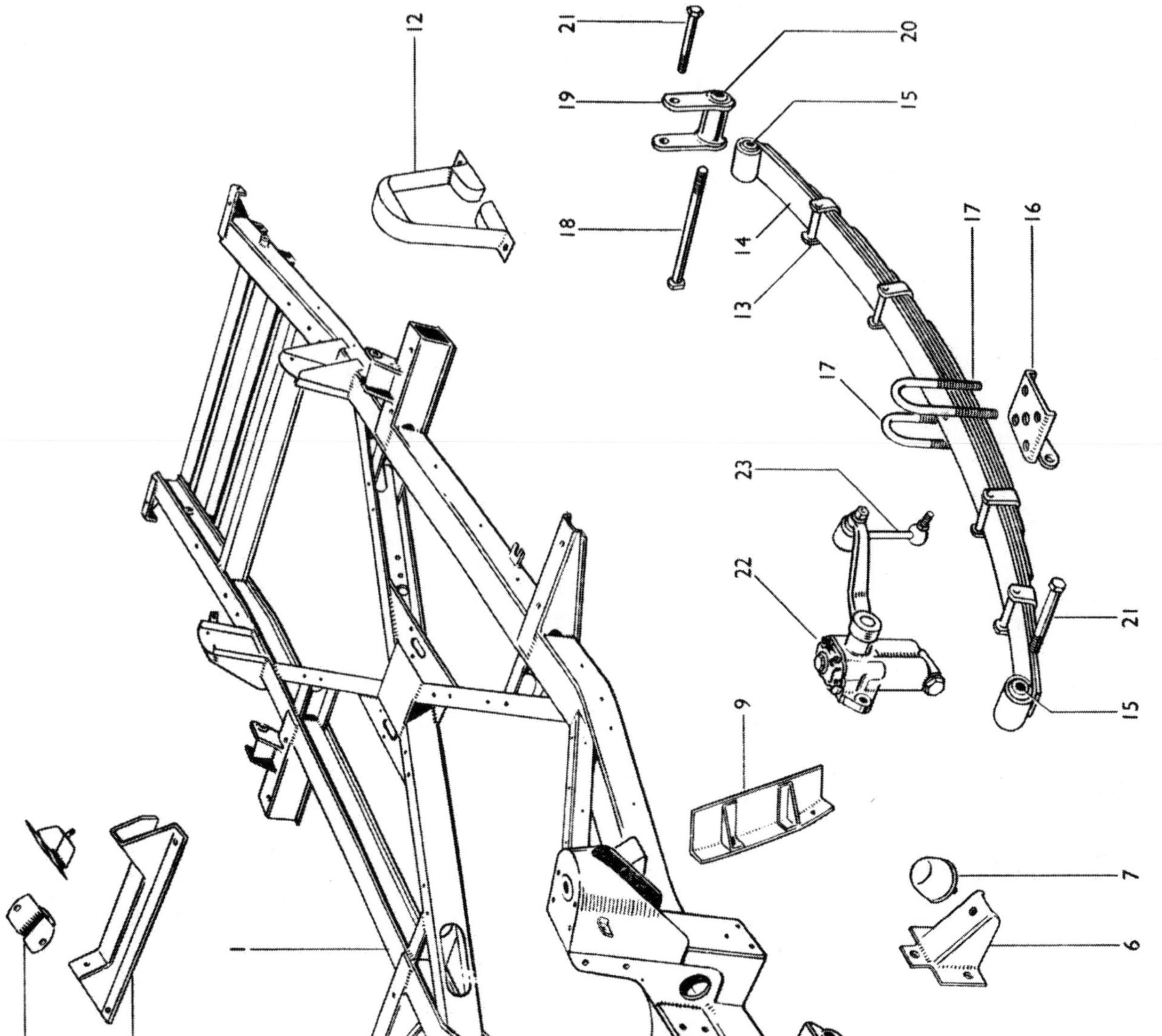

PLATE 27

Part No.	Description	Plate No.	No. per Unit	Remarks
26266	**CHASSIS FRAME ASSEMBLY**	27—1	1	
307773	**CLAMPING BRACKET FOR STEERING COLUMN AND UNIT**	27—2	1	
UFB.137/52R	**Bolt,** clamping Steering Column		1	
C.8867/3	**Nut,** Nyloc, on Bolt		1	
UFB.137/22R	**Bolt,** securing Bracket to Chassis		2	
C.8667/3	**Nut,** Nyloc, on Bolts		2	
136051	**BRACING MEMBER FOR FRONT OF CHASSIS FRAME**	27—3	1	
UFS.137/8R	**Screw,** Set, securing Member to Chassis		4	
FG.106/X	**Washer,** Spring, on Setscrews		4	
307598	**ENGINE FRONT SUPPORT BRACKET**	27—4	2	
307129	**ENGINE MOUNTING RUBBER (METALASTIK)**	27—5	2	
307578	**FRONT SUSPENSION REBOUND STOP BRACKET**	27—6	2	
UFS.137/7R	**Screw,** Set, securing Brackets to Chassis		6	
FG.106/X	**Washer,** Spring, on Setscrews		6	
431458	**RUBBER REBOUND STOP ASSEMBLY**	27—7	2	
C.8667/3	**Nut,** Nyloc, securing Rebound Stops to Chassis		2	
308125	**R.H. MUDSHIELD**	27—8	1	
308126	**L.H. MUDSHIELD**	27—9	1	
308127	**PETROL PIPE SHIELD**		1	
423196/43	**Screw,** Self-Tapping securing Shields		8	
136200	**GEARBOX SUPPORT BRACKET**	27—10	1	
UFS.131/7R	**Screw,** Set, securing Bracket to Chassis		4	
C.8667/2	**Nut,** Nyloc, on Bolts		4	
FW.105/T	**Washer,** Plain, under Nuts		4	

Part No.	Description	Plate No.	No. per Unit	Remarks
307497	**GEARBOX MOUNTING RUBBER**	27—11	2	
C.8667/2	**Nut,** securing Mounting Rubber to Bracket		2	
UCS.131/5R	**Screw,** Set, securing Mounting Rubber to Gearbox		4	NOT REQUIRED WHEN AUTOMATIC TRANSMISSION IS FITTED
FG.105/X	**Washer,** Spring, on Setscrews		4	
308968	**REAR AXLE CHECK STRAP ASSEMBLY**	27—12	2	
UFS.137/6R	**Screw,** Set, securing Check Strap to Chassis		4	
FG.106/X	**Washer,** Spring, on Setscrews		4	
430521/05	**REAR ROAD SPRING**	27—13	2	
430531	**Main Leaf Assembly**		2	
430531	**MAIN LEAF ASSEMBLY**	27—14	2	
430532	**Bush,** in Main Leaf	27—15	4	
431322	**Bottom Plate,** for R.H. Rear Spring		1	
431323	**Bottom Plate,** for L.H. Rear Spring	27—16	1	
431320	**'U' Bolt,** securing Rear Road Springs to Rear Axle	27—17	4	
C.8667/4	**Nut,** Nyloc, on 'U' Bolts		8	
431313	**Bolt,** securing rear end of Springs to Shackle	27—18	2	
C.8667/7	**Nut,** Nyloc, on Bolts		2	
FW.110/T	**Washer,** Plain, under Nuts		2	
431314	**REAR ROAD SPRING SHACKLE ASSEMBLY**	27—19	2	
430701	**Bush,** in Shackle	27—20	2	
UFB.150/30R	**Bolt,** securing Shackles to Chassis	27—21	4	
C.8667/5	**Nut,** Nyloc, on Bolts		4	
307641/01	**R.H. REAR SHOCK ABSORBER**	27—22	1	
307642/01	**L.H. REAR SHOCK ABSORBER**		1	
UFB.143/11R	**Bolt,** securing Shock Absorbers to Chassis		4	
C.8667/4	**Nut,** Nyloc, on Bolts		4	
307641/02	**LINK, CONNECTING SHOCK ABSORBER TO BOTTOM PLATE ON ROAD SPRING**	27—23	2	
C.8667/4	**Nut,** securing Link to Shock Absorber		2	
C.8667/5	**Nut,** securing Link to Bottom Plate		2	

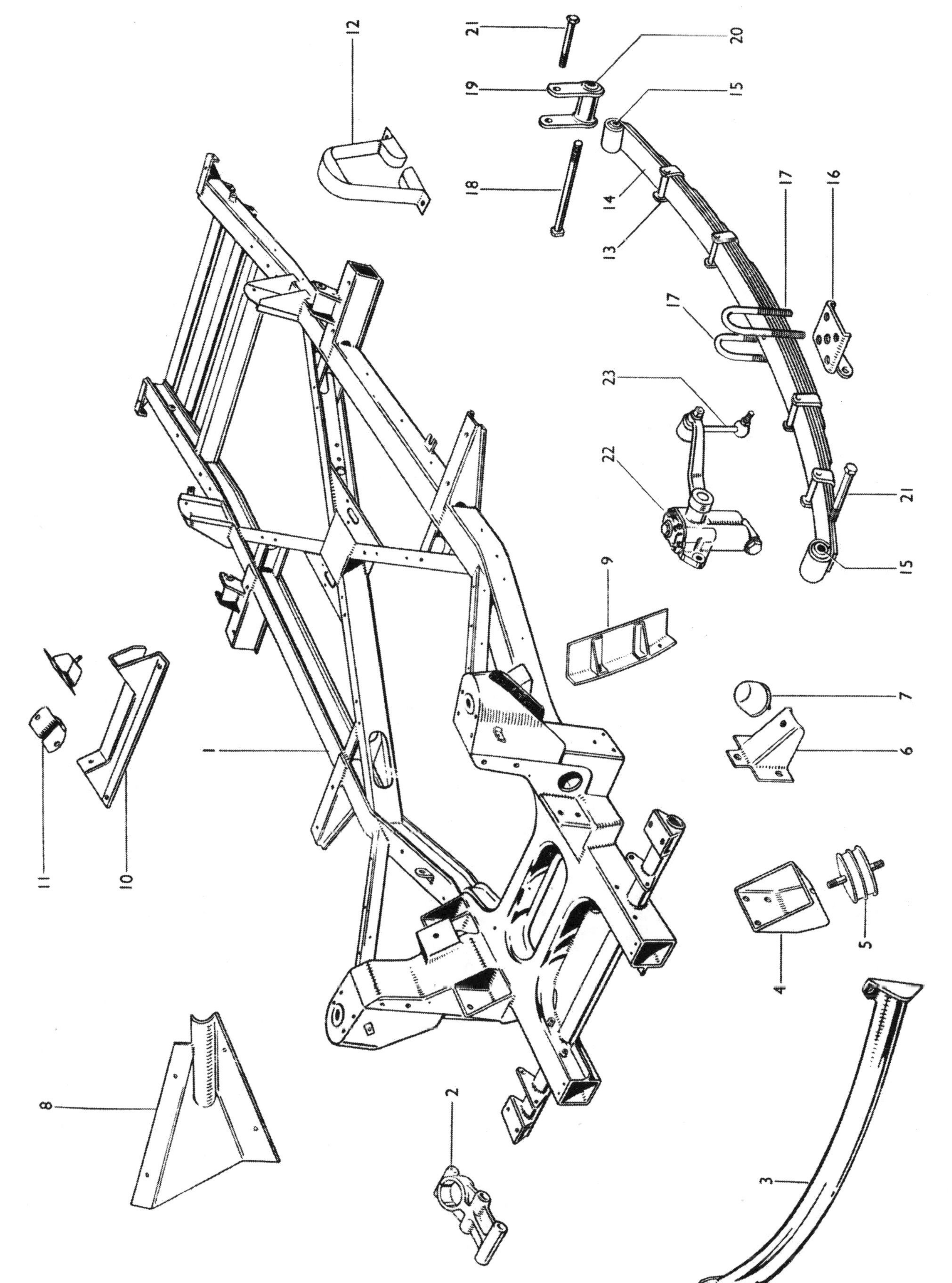

PLATE 27

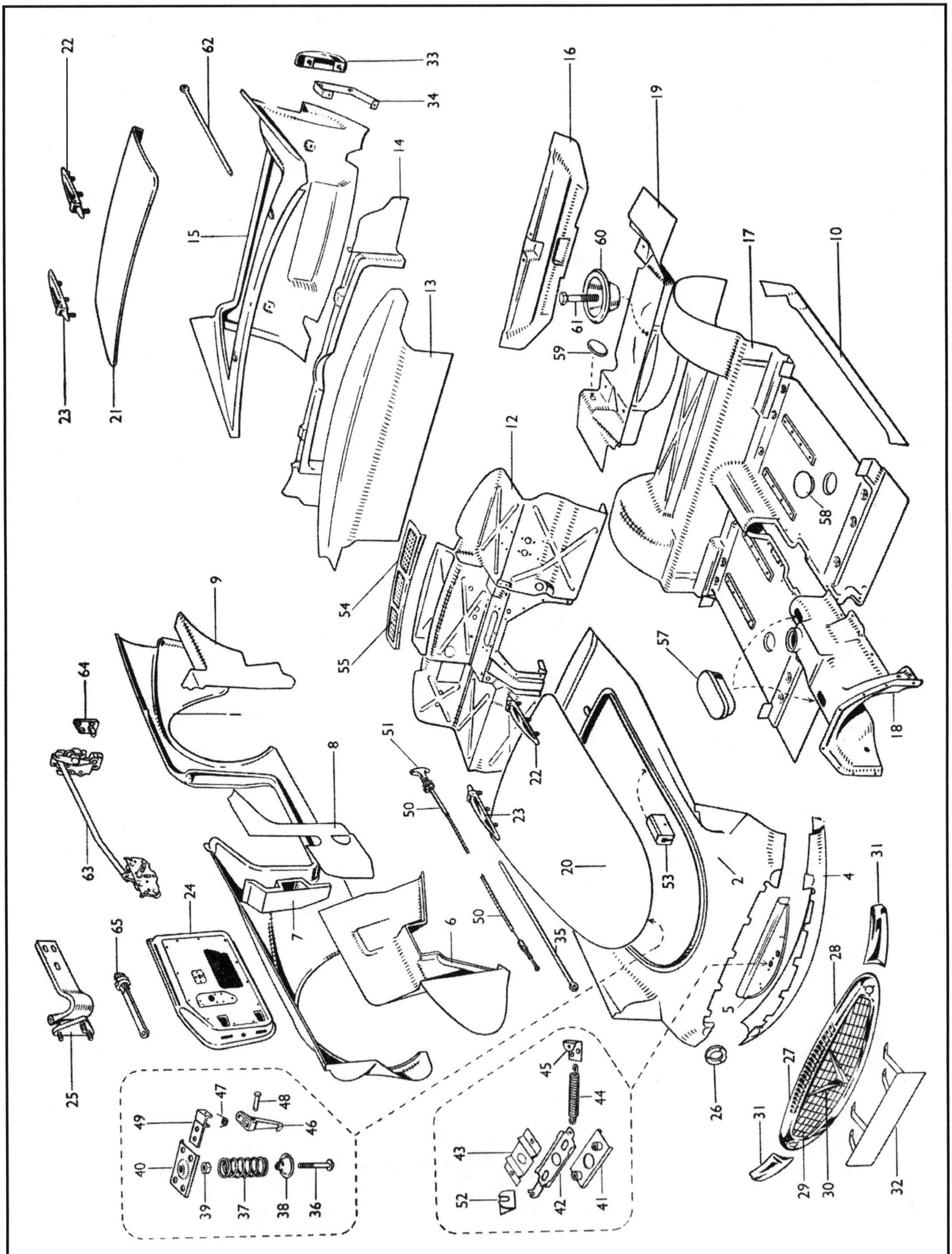

PLATE 28

Part No.	Description	Plate No.	No. per Unit	Remarks
81/621013	**BODY SHELL ASSEMBLY**		1	**R.H. DRIVE CARS ONLY**
81/692979	**Body Side Assembly,** L.H.		1	
81/692980	**Body Side Assembly,** R.H.	28—1	1	
81/699150	**Bonnet Aperture Panel Assembly**	28—2	1	
81/669148	**Body Front Lower Panel Assembly**	28—4	1	
81/669121	**Bonnet Lock Panel Assembly**	28—5	1	
81/669079	**Valance Panel Assembly,** L.H.		1	
81/669080	**Valance Panel Assembly,** R.H.	28—6	1	
81/625249	**'A' Post Reinforcement,** L.H.		1	**1st CONDITION**
81/625250	**'A' Post Reinforcement,** R.H.	28—7	1	**1st CONDITION**
170745	**'A' Post Reinforcement,** L.H.		1	**2nd CONDITION**
170746	**'A' Post Reinforcement,** R.H.		1	**2nd CONDITION**
81/615131	**'A' Post Closing Panel,** L.H.		1	
81/615132	**'A' Post Closing Panel,** R.H.	28—8	1	
81/622173	**'B' Post Closing Panel,** L.H.		1	
81/622174	**'B' Post Closing Panel,** R.H.	28—9	1	
81/614051	**Sill Reinforcement,** L.H.	28—10	1	
81/614052	**Sill Reinforcement,** R.H.		1	
81/625008	**Dash Assembly**	28—12	1	
81/622269	**Rear Squab Panel Assembly**	28—13	1	
81/622290	**Tonneau Panel/Boot Hinge Reinforcement Assembly**	28—14	1	
81/692983	**Boot Aperture Panel**	28—15	1	
81/622295	**Body Rear Lower Outer Panel Assembly**	28—15	1	
81/622000	**Body Rear Lower Inner Panel Assembly**	28—16	1	
81/624004	**Main Floor Assembly**	28—17	1	
81/644650	**Gearbox Cover Assembly**	28—18	1	**NOT REQUIRED WHEN AUTOMATIC TRANSMISSION IS FITTED**
53015	**Gearbox Cover Assembly**		1	**REQUIRED WHEN AUTOMATIC TRANSMISSION IS FITTED**
81/624098	**Luggage Floor and Spare Wheel Pan Assembly**	28—19	1	
82/621013	**BODY SHELL ASSEMBLY**		1	**L.H. DRIVE CARS ONLY**
82/669079	**Valance Panel Assembly,** L.H.		1	
82/669080	**Valance Panel Assembly,** R.H.		1	
82/625008	**Dash Assembly**	28—12	1	
	All other items are as for Body Shell Assembly 81/621013			
81/669001	**BONNET LID ASSEMBLY**	28—20	1	
81/626004	**BOOT LID ASSEMBLY**	28—21	1	
65/669043	**HINGE ASSEMBLY, L.H. FOR BONNET AND BOOT LIDS**	28—22	2	
65/669044	**HINGE ASSEMBLY, R.H. FOR BONNET AND BOOT LIDS**	28—23	2	
C.8667/1	**Nut,** Nyloc, securing Hinges to Bonnet and Boot Lids		8	
FW.104/T	**Washer,** Plain, under Nuts		8	
C.8667/2	**Nut,** Nyloc, securing Hinges to Body		4	
FW.105/T	**Washer,** Plain, under Nuts		4	

BODY DETAILS

Part No.	Description	Plate No.	No. per Unit	Remarks
81/626013	**L.H. DOOR ASSEMBLY**		1	**1st CONDITION**
81/626014	**R.H. DOOR ASSEMBLY**		1	
81/626240	**L.H. DOOR HINGE ASSEMBLY**		2	
81/626241	**R.H. DOOR HINGE ASSEMBLY**		2	**Note:—These Doors are identified by a recess behind the exterior handle**
81/616300	**Hinge Plate,** at 'A' Posts		4	
UFB.131/8R	**Bolt,** securing Hinges to Doors		12	
C.725	**Washer,** Shakeproof, on Bolts		12	
FW.105/T	**Washer,** Plain, under Shakeproof Washers		12	
C.8667/2	**Nut,** Nyloc, securing Hinges to Body		12	
53021	**L.H. DOOR ASSEMBLY**		1	**2nd CONDITION**
53022	**R.H. DOOR ASSEMBLY**	28—24	1	
370024	**L.H. DOOR HINGE ASSEMBLY**		2	
370025	**R.H. DOOR HINGE ASSEMBLY**	28—25	2	
UFB.131/8R	**Bolt,** securing Hinges to Doors		12	
C.725	**Washer,** Shakeproof, on Bolts		12	
FW.105/T	**Washer,** Plain, under Shakeproof Washers		12	
87536/10	**Screw,** Phillips Head, securing Hinges to Body		12	
470004	**DOOR SILENCER (FEMALE)**		2	
431462/36	**Screw,** Set, securing Door Silencers to Body		6	
470010A	**DOOR SILENCER (MALE)**		2	
431462/36	**Screw,** Set, securing Door Silencers to Doors		4	
81/659055	**Tapping Ring,** for Front Flasher and Rear Lamps	28—26	6	
26222	**FRONT GRILLE ASSEMBLY**	28—27	1	
26222/01	**Surround,** for Front Grille	28—28	1	
26222/02	**Mesh,** for Front Grille	28—29	1	
26222/03	**Nut,** securing Mesh		10	
26222/04	**'V' Motif,** on Grille	28—30	1	

Part No.	Description	Plate No.	No. per Unit	Remarks
	FRONT GRILLE ASSEMBLY (Cont.)			
26222/05	**Retainer,** securing 'V' Motif		2	
26222/06	**Strap,** fixing Motif		2	
26222/07	**Strap,** fixing Mesh		20	
C.8667/1	**Nut,** Nyloc, securing Front Grille to Body		12	
51/637052	**Washer,** Plain, under Nuts		12	
431912/1	**MOULDING (P.V.C.) BETWEEN FRONT GRILLE AND BODY (89″ LONG)**		1	
136190	**FRONT BUMPER, AT EACH SIDE OF FRONT GRILLE**	28—31	2	1st CONDITION
UFS.137/7R	**Screw,** Set, securing Bumpers to Body		4	1st CONDITION
FG.106/X	**Washer,** Spring, on Setscrews		4	1st CONDITION
FW.106/T	**Washer,** Plain, under Spring Washers		4	1st CONDITION
431912/2	**MOULDING (P.V.C.) BETWEEN BUMPER AND BODY (16″ LONG)**		2	1st CONDITION
136351	**FRONT BUMPER ASSEMBLY (FULL LENGTH)**		1	2nd CONDITION
136256	**Bumper Bar Assembly**		1	2nd CONDITION
308343	**R.H. Over-rider Assembly**		1	2nd CONDITION
308344	**L.H. Over-rider Assembly**		1	2nd CONDITION
431912/3	**Over-rider Moulding** (P.V.C.) (7″ long)		4	2nd CONDITION
431876	**Distance Piece,** for Over-riders		2	2nd CONDITION
UFN.137/L	**Nut,** securing Over-riders		2	2nd CONDITION
FG.106/X	**Washer,** Spring, under Nuts		2	2nd CONDITION
136260	**Spring Bar,** supporting Bumper Bar		1	2nd CONDITION
UFB.137/7R	**Bolt,** securing Spring Bar to Bumper		3	2nd CONDITION
FG.106/X	**Washer,** Spring, on Bolts		3	2nd CONDITION
FW.106/T	**Washer,** Plain, under Spring Washers		3	2nd CONDITION
136262	**BRACKET, R.H., SECURING FRONT BUMPER TO CHASSIS**		1	2nd CONDITION
136263	**BRACKET, L.H., SECURING FRONT BUMPER TO CHASSIS**		1	2nd CONDITION
UFB.137/24R	**Bolt,** securing Brackets to Chassis		4	2nd CONDITION
UFN.137/L	**Nut,** on Bolts		4	2nd CONDITION
FG.106/X	**Washer,** Spring, under Nuts		4	2nd CONDITION
UFB.137/6R	**Bolt,** securing Bumper Assembly to Brackets		2	2nd CONDITION
UFN.137/L	**Nut,** on Bolts		2	2nd CONDITION
FG.106/X	**Washer,** Spring, under Nuts		2	2nd CONDITION
308159	**FRONT NUMBER PLATE SUPPORT**	28—32	1	1st CONDITION
UFS.125/4R	**Screw,** Set, securing Support to Body		4	1st CONDITION
FG.104/X	**Washer,** Spring, on Setscrews		4	1st CONDITION
FW.104/T	**Washer,** Plain, under Spring Washers		4	1st CONDITION

BODY DETAILS

Part No.	Description	Plate No.	No. per Unit	Remarks
308332	**FRONT NUMBER PLATE SUPPORT**		1	2nd CONDITION
UFS.125/5R	**Screw,** Set, securing Support to Front Bumper		2	
C.8667/1	**Nut,** Nyloc, on Setscrews		2	
53/637073	**MOTIF, "DAIMLER, ENGLAND"**		1	
53/637074	**Clip,** securing Motif to Body		2	
136189	**REAR OVER-RIDER ASSEMBLY**	28—33	2	
431491	**Bolt,** Special, securing Over-rider to Body		4	
UFN.137/L	**Nut,** on Bolts		4	
FG.106/X	**Washer,** Spring, under Nuts		4	
FW.106/T	**Washer,** Plain, under Spring Washer		4	
307997	**BRACKET MOUNTING REAR OVER-RIDERS**	28—34	2	
UFS.137/8R	**Screw,** Set, securing Bracket to Chassis		2	
FG.106/X	**Washer,** Spring, on Setscrews		2	
FW.106/T	**Washer,** Plain, under Spring Washers		2	
UFS.137/10R	**Screw,** Set, securing Bracket to Body		2	
C.8667/3	**Nut,** Nyloc, on Setscrews		2	
FW.106/T	**Washer,** Plain, under Nuts		2	
81/659184	**BONNET PROP ROD**	28—35	1	
422531/03	**Clevis Pin,** pivoting Bonnet Prop.		1	
L.102/5U	**Split Pin,** retaining Clevis Pin		1	
426558	**Clip,** retaining Bonnet Prop when not in use		1	
425291/25	**Screw,** securing Clip to Body		1	
81/669204	**BONNET LOCK STRIKER ASSEMBLY**		1	1st CONDITION
81/659221	**Striker Pin**	28—36	1	
81/659220	**Spring,** on Striker Pin	28—37	1	
81/669222	**Retainer,** for Spring	28—38	1	
900003	**Nut,** securing Striker Pin	28—39	1	
81/669219	**Striker Pin Plate**	28—40	1	
92/669204	**BONNET LOCK STRIKER ASSEMBLY**		1	2nd CONDITION
81/659222	**Retainer,** for Spring		1	
92/669219	**Striker Pin Plate**		1	
	All other items as for Bonnet Lock Striker Assembly 81/669204			
UFS.125/4R	**Screw,** Set, securing Striker Plate to Bonnet		4	
C.724	**Washer,** Shakeproof, on Setscrews		4	
FW.104/T	**Washer,** Plain, under Shakeproof Washers		4	

Part No.	Description	Plate No.	No. per Unit	Remarks
92/669201	**BONNET LOCK STRIKER PLATE ASSEMBLY**	28—41	1	
92/659202	**Slider,** for Bonnet Lock	28—42	1	
81/659218	**Locating Plate,** for Slider	28—43	1	
BD.12855	**Spring,** tensioning Slider	28—44	1	
81/659226	**Anchor Bracket,** for Spring	28—45	1	
423195/34	**Screw,** Set, securing Anchor Bracket		2	
11404/10	**Screw,** Set, securing Bonnet Lock Striker Plate Assembly to Body		2	
C.724	**Washer,** Shakeproof, on Setscrews		2	
FW.104/T	**Washer,** Plain, under Shakeproof Washers		2	
51/637052	**Washer,** Special, on Setscrews		2	
81/669213	**BONNET SAFETY CATCH ASSEMBLY**		1	
81/659214	**Safety Hook,** for Bonnet	28—46	1	
81/659215	**Spring,** tensioning Safety Hook	28—47	1	
81/659216	**Clevis Pin,** pivoting Safety Hook	28—48	1	
L.102/3U	**Split Pin,** retaining Clevis Pin		1	
81/669116	**Bracket,** mounting Safety Hook	28—49	1	**1st CONDITION**
81/669158	**Bracket,** mounting Safety Hook		1	**2nd CONDITION**
308204	**BONNET LOCK RELEASE CABLE**	28—50	1	
431621	**Handle,** on Release Cable	28—51	1	
431622	**Grubscrew,** securing Handle to Cable		1	
81/659225	**ABUTMENT BRACKET FOR BONNET LOCK RELEASE CABLE**	28—52	1	
423195/34	**Screw,** Set, securing Bracket to Body		2	
81/659094	**BONNET TOP RUBBER BUFFER**	28—53	6	
423192/11	**Rivet,** securing Buffers to Body		6	
81/641105	**PLENUM CHAMBER GRILLE ASSEMBLY**	28—54	1	
90012/04	**Screw,** securing Grille Assembly to Body		4	
81/631106	**PLENUM CHAMBER LATTICE**	28—55	1	
75/637467	**RUBBER PLUG IN GEARBOX COVER**	28—57	2	**1 OFF ONLY REQUIRED WHEN AUTOMATIC TRANSMISSION IS FITTED**
81/637213	**BLANKING PLUG IN GEARBOX COVER**		1	**REQUIRED ONLY WHEN AUTOMATIC TRANSMISSION IS FITTED**
81/637213	**BLANKING PLUG, IN MAIN FLOOR**	28—58	2	

Part No.	Description	Plate No.	No. per Unit	Remarks
81/637187	**BLANKING PLUG, IN SPARE WHEEL FLOOR**	28—59	2	
81/644190	**SPARE WHEEL CLAMP ASSEMBLY**	28—60	1	
87625	**Bolt,** Special, for Spare Wheel Clamp	28—61	1	
81/632175	**BOOT LID PROP ROD**	28—62	1	
422531/03	**Clevis Pin,** pivoting Prop Rod		1	
81/637087	**Wave Washer,** on Clevis Pin		1	
L.102/5U	**Split Pin,** retaining Clevis Pin		1	
FW.104/T	**Washer, Plain,** behind Split Pin		1	
426588	**Clip,** retaining Prop Rod when not in use		1	
425291/25	**Screw,** securing Clip to Body		1	
92/659145	**BOOT LID RUBBER BUFFER**		2	
81/632598	**BOOT LID SEALING RUBBER (138" LONG)**		1	
81/626061	**L.H. DOOR LOCK AND REMOTE CONTROL ASSEMBLY**		1	
81/626062	**R.H. DOOR LOCK AND REMOTE CONTROL ASSEMBLY**	28—63	1	
NS.419/3H	**Screw,** securing Door Lock and Remote Control to Doors		12	
AW.102/T	**Washer,** Plain, on Screws		12	
NS.419/4H	**Safety Screw,** for Door Lock (Locking)		1	
81/626521	**STRIKER PLATE, FOR L.H. DOOR LOCK**		1	
81/626522	**STRIKER PLATE, FOR R.H. DOOR LOCK**	28—64	1	
UFS.525/7H	**Screw,** securing Striker Plates to Body		6	
81/621054	**DOOR CHECK STRAP ASSEMBLY**	28—65	2	
51/617089	**Rivet,** securing Check Strap to 'A' Post		2	
81/641087	**WINDOW REGULATOR IN L.H. DOOR**		1	

Part No.	Description	Plate No.	No. per Unit	Remarks
81/641088	**WINDOW REGULATOR IN R.H. DOOR**	29—66	1	
81/646238	**Support Panel,** for Window Regulator	29—67	2	
423197/01	**Screw,** securing Window Regulators to Support Panels		12	
422543/1	**Nut,** on Screws		12	
422508/12	**Washer,** Shakeproof, under Nuts		12	
NS.419/4H	**Screw,** securing Regulators and Panels to Doors		12	
422508/09	**Washer,** Shakeproof, on Screws		12	
81/636242	**HANDLE, OPERATING DOOR WINDOW REGULATORS**	29—68	2	
51/636245	**Escutcheon,** for Handles	29—69	2	
51/636244	**Locking Pin,** retaining Handles		2	
81/646445	**L.H. DOOR GLASS FRONT CHANNEL**		1	
81/646446	**R.H. DOOR GLASS FRONT CHANNEL**	29—70	1	
81/636375	**Felt,** for Front Channels		2	
UFS.425/6H	**Screw,** securing Front Channels to Doors		4	
51/637052	**Spacing Washer,** on Screws		As req'd	
81/646447	**L.H. DOOR GLASS REAR CHANNEL**		1	
81/646448	**R.H. DOOR GLASS REAR CHANNEL**	29—71	1	
81/636376	**Felt,** for Rear Channels		2	
UFS.425/5H	**Screw,** securing Rear Channels to Doors		2	
UFS.425/6H	**Screw,** securing Rear Channels to Doors		2	
FW.104/T	**Washer,** Plain, on Screws		4	
51/637052	**Spacing Washer,** on Screws		As req'd	
81/646123	**L.H. DOOR GLASS BOTTOM CHANNEL**		1	
81/646124	**R.H. DOOR GLASS BOTTOM CHANNEL**	29—72	1	
81/636112	**Rubber,** in Bottom Channels	29—73	2	
81/646215	**DOOR GLASS AND FRAME ASSEMBLY**	29—74	2	
81/646128	**DOOR HANDLE, EXTERIOR (NON-LOCKING)**	29—75	1	
81/646592	**Push Button Assembly,** for Door Handle	29—76	1	

BODY DETAILS

Part No.	Description	Plate No.	No. per Unit	Remarks
81/646127	**DOOR HANDLE, EXTERIOR (LOCKING)**		1	
81/646593	**Push Button Assembly,** for Door Handle		1	
81/636197	**Lock Barrel,** for Door Handle		1	
81/636188	**Seating Washer** (Large) for Door Handles	29—77	2	
81/636189	**Seating Washer** (Small) for Door Handles	29—78	2	
900337	**Screw,** securing Exterior Handles to Doors		2	
C.724	**Washer,** Shakeproof, on Screws		2	
FW.104/T	**Washer,** Plain, under Shakeproof Washers		2	
423195/33	**Screw,** securing Exterior Handles to Doors		2	
81/636243	**DOOR HANDLE, INTERIOR**	29—79	2	
51/636245	**Escutcheon,** under Handles	29—80	2	
51/636244	**Locking Pin,** retaining Handles		2	
81/636464	**SEALING RUBBER FOR DOORS**		2	
81/632513	**DRAUGHT EXCLUDER FOR DOORS**		2	
81/645226	**WINDSCREEN ASSEMBLY (TOUGHENED GLASS)**	29—81	1	
81/635100	**Glass,** for Windscreen (Toughened)	29—82	1	
81/635257	**Top Frame,** for Windscreen	29—83	1	
81/635258	**Bottom Frame,** for Windscreen	29—84	1	
81/635255	**L.H. Frame,** for Windscreen	29—85	1	
81/635256	**R.H. Frame,** for Windscreen	29—86	1	
81/635378	**Rubber,** Glazing for Windscreen	29—87	1	
81/635234	**Rubber,** Sealing between Windscreen and Scuttle	29—88	1	
81/635254	**Rubber,** Sealing between Windscreen and Roof	29—89	1	
81/675226	**WINDSCREEN ASSEMBLY (LAMINATED GLASS)**		1	
81/675130	**Glass,** for Windscreen (Laminated)		1	
	All other items are as for Windscreen Assembly, 81/645226			
C.8667/2	**Nut,** Nyloc, securing Windscreen to Body		3	
FW.105/T	**Washer,** Plain, under Nuts		3	
87561	**Shouldered Bolt,** securing Windscreen to Body		2	
FG.104/X	**Washer,** Spring, on Shouldered Bolt		2	
FW.104/T	**Washer,** Plain, under Spring Washer		2	
81/635315	**L.H. SEALING RUBBER FOR WINDSCREEN PILLAR**		1	
81/635316	**R.H. SEALING RUBBER FOR WINDSCREEN PILLAR**		1	

Part No.	Description	Plate No.	No. per Unit	Remarks
53227	**FACIA PANEL ASSEMBLY (FULLY TRIMMED)**		1	R.H. DRIVE CARS ONLY
81/641169	**Cubby Box,** in Facia Panel		1	R.H. DRIVE CARS ONLY
81/645151	**Lid,** for Cubby Box		1	R.H. DRIVE CARS ONLY
53228	**FACIA PANEL ASSEMBLY (FULLY TRIMMED)**	29—90	1	L.H. DRIVE CARS ONLY
82/641170	**Cubby Box,** in Facia Panel		1	L.H. DRIVE CARS ONLY
82/645152	**Lid,** for Cubby Box		1	L.H. DRIVE CARS ONLY
UFS.525/6H	**Screw,** securing Facia Panel to Body		4	
81/645151	**CUBBY BOX LID (TRIMMED)**		1	R.H. DRIVE CARS ONLY
92/637380	**Hinge,** for Cubby Box Lid		2	R.H. DRIVE CARS ONLY
82/645152	**CUBBY BOX LID (TRIMMED)**		1	L.H. DRIVE CARS ONLY
92/637380	**Hinge,** for Cubby Box Lid		2	L.H. DRIVE CARS ONLY
53/645133	**LOCK FOR CUBBY BOX LID**		1	
425282/21	**Screw,** securing Lock to Lid		2	
425281/02	**Screw,** securing Lock to Lid		1	
53/645134	**Striker,** for Cubby Box Lid Lock		1	
425282/20	**Screw,** securing Striker to Facia Panel		2	
53225	**FACIA TRIM PANEL ASSEMBLY**		1	R.H. DRIVE CARS ONLY
53226	**FACIA TRIM PANEL ASSEMBLY**	29—91	1	L.H. DRIVE CARS ONLY
81/637044	**Clip,** securing Trim Panel to Facia		10	
81/645500	**CRASH RAIL ASSEMBLY (TRIMMED)**	29—92	1	
NN.119/L	**Nut,** securing Crash Rail to Body		7	
AW.102/T	**Washer,** Plain, under Nuts		7	
81/641799	**INTERIOR DRIVING MIRROR**	29—93	1	
425281/33	**Screw,** securing Mirror		2	
81/647063	**ASHTRAY IN GEARBOX COVER**		1	

BODY DETAILS

Part No.	Description	Plate No.	No. per Unit	Remarks
53108	**FRONT SEAT ASSEMBLY (FULLY TRIMMED)**		2	
81/644028	**Frame Assembly**	29—94	2	
81/634091	**Slide** (Plain) at L.H. side of Front Seats	29—95	2	
81/634092	**Slide** (Catch) at R.H. side of Front Seats	29—96	2	
UFB.125/10R	**Bolt,** securing Slides to Seat Frames		6	
UFB.125/12R	**Bolt,** securing Catch Slides to Seat Frames		2	
UFN.125/L	**Nut,** Nyloc, on Bolts		8	
81/637541	**Clip,** securing Seat Trim to Seat Frames		32	
81/644093	**L.H. Floor Rail,** for Seats	29—97	2	
81/644094	**R.H. Floor Rail,** for Seats	29—98	2	
81/644552	**HANDBRAKE LEVER LEATHER GAITER**		1	
81/641840	**PETROL TANK REAR SUPPORT STRAP ASSEMBLY**	29—99	2	
81/641839	**PETROL TANK TOP SUPPORT STRAP ASSEMBLY**	29—100	2	
UFS.125/4R	**Screw,** Set, securing Strap (Rear to Top)		2	
UFN.125/L	**Nut,** Nyloc, on Setscrews		2	
UFS.125/7R	**Screw,** Set, securing Straps to Body		4	
C.724	**Washer,** Shakeproof, on Setscrews		4	
81/646187	**BOOT LID LOCK ASSEMBLY**	29—101	1	
81/646802	**Push Button Assembly**	29—102	1	
81/636800	**Locking Barrel**		1	
81/637503	**Key,** for Locking Barrel	29—103	1	
UFS.125/4R	**Screw,** Set, securing Boot Lid Lock to Body		2	
C.724	**Washer,** Shakeproof, on Setscrews		2	
81/631500	**Striker,** for Boot Lid Lock	29—104	1	
423195/34	**Screw,** securing Striker to Boot Lid		2	
81/646227	**L.H. DOOR CASING (TRIMMED)**		1	1st CONDITION
81/646228	**R.H. DOOR CASING (TRIMMED)**		1	1st CONDITION
170138	**L.H. DOOR CASING (TRIMMED)**		1	2nd CONDITION
170139	**R.H. DOOR CASING (TRIMMED)**		1	2nd CONDITION
81/645123	**L.H. DASH SIDE CASING (TRIMMED)**		1	L.H. DRIVE CARS ONLY 1st CONDITION
81/645124	**R.H. DASH SIDE CASING (TRIMMED)**		1	L.H. DRIVE CARS ONLY 1st CONDITION

Part No.	Description	Plate No.	No. per Unit	Remarks
170227	**L.H. DASH SIDE CASING (TRIMMED)**		1	L.H. DRIVE CARS ONLY 2nd CONDITION
170228	**R.H. DASH SIDE CASING (TRIMMED)**		1	
82/645123	**L.H. DASH SIDE CASING (TRIMMED)**		1	R.H. DRIVE CARS ONLY 1st CONDITION
82/645124	**R.H. DASH SIDE CASING (TRIMMED)**		1	
170225	**L.H. DASH SIDE CASING (TRIMMED)**		1	R.H. DRIVE CARS ONLY 2nd CONDITION
170226	**R.H. DASH SIDE CASING (TRIMMED)**		1	
81/642119	**L.H. REAR QUARTER CASING**		1	1st CONDITION
81/642120	**R.H. REAR QUARTER CASING**		1	
170237	**L.H. REAR QUARTER CASING**		1	2nd CONDITION
170238	**R.H. REAR QUARTER CASING**		1	
81/643195	**HOOD ASSEMBLY (COMPLETE)**		1	
81/643177	**HOOD MECHANISM ASSEMBLY**		1	
81/643190	**CANOPY RAIL ASSEMBLY**		1	
81/643180	**Toggle Fastener,** securing Hood to Windscreen		3	
81/633407	**Sealing Rubber,** between Canopy and Windscreen		1	
81/643229	**HOOD COVER ASSEMBLY (COMPLETE WITH BACKLIGHT)**		1	
80480	**Button Fastener,** securing Hood Cover to Body (Tenax)		19	
80469	**Nut,** securing Button Fasteners		19	
80466	**Eyelet,** in Hood Cover for Turnbuttons		2	
80467	**Washer,** for Eyelets		2	

BODY DETAILS

Part No.	Description	Plate No.	No. per Unit	Remarks
81/643441	**HOOD ENVELOPE**		1	
80480	**Button Fastener,** securing Hood Envelope to Body (Tenax)		19	
80469	**Nut,** securing Button Fasteners		19	
80470	**Dot Fastener Button**		2	
80472	**Dot Fastener Socket**		2	
80466	**Eyelet,** in Hood Envelope for Turnbuttons		2	
80467	**Washer,** for Eyelets		2	
80468	**Stud,** receiving Button Fasteners		2	
470035	**Stud,** receiving Button Fasteners		17	
88628	**Nut,** securing Studs		19	
80465	**Turnbutton Fastener,** securing Hood Cover or Envelope		2	
80473	**Stud,** for Dot Fasteners		2	
423190/13	**Screw,** securing Studs		2	
81/644457	**L.H. FRONT FLOOR CARPET ASSEMBLY**		1	**R.H. DRIVE CARS ONLY 1st CONDITION**
81/644458	**R.H. FRONT FLOOR CARPET ASSEMBLY**		1	
53058	**L.H. FRONT FLOOR CARPET ASSEMBLY**		1	**R.H. DRIVE CARS ONLY 2nd CONDITION**
53059	**R.H. FRONT FLOOR CARPET ASSEMBLY**		1	
82/644457	**L.H. FRONT FLOOR CARPET ASSEMBLY**		1	**L.H. DRIVE CARS ONLY 1st CONDITION**
82/644458	**R.H. FRONT FLOOR CARPET ASSEMBLY**		1	
53062	**L.H. FRONT FLOOR CARPET ASSEMBLY**		1	**L.H. DRIVE CARS ONLY 2nd CONDITION**
53063	**R.H. FRONT FLOOR CARPET ASSEMBLY**		1	
81/644753	**INTERMEDIATE FLOOR CARPET ASSEMBLY**		2	
81/645512	**DASH CENTRE CARPET ASSEMBLY**		1	**R.H. DRIVE CARS ONLY**

Part No.	Description	Plate No.	No. per Unit	Remarks
82/645512	DASH CENTRE CARPET ASSEMBLY		1	L.H. DRIVE CARS ONLY
81/644450	GEARBOX COVER CARPET		1	
81/644455	PROPELLOR SHAFT TUNNEL CARPET		1	

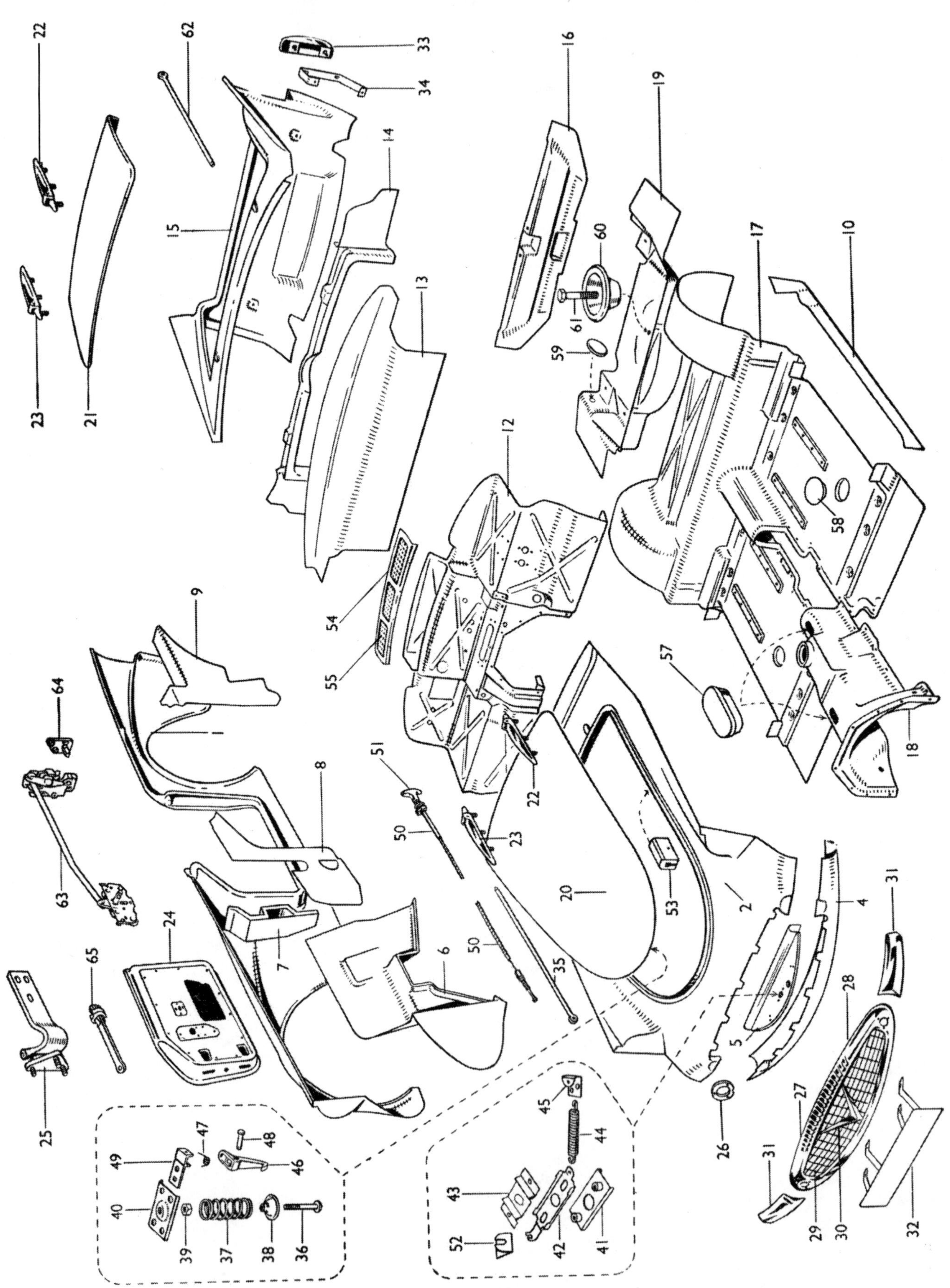

22
62
33
34
16
19
23
21
15
14
13
60
17
10
61
59
12
58
9
64
54
55
57
18
51
8
22
50
23
63
20
53
2
24
65
7
6
50
35
31
4
28
5
25
47
49
48
46
40
39
37
38
36
45
44
43
52
42
41
26
27
31
29
30
32

PLATE 28

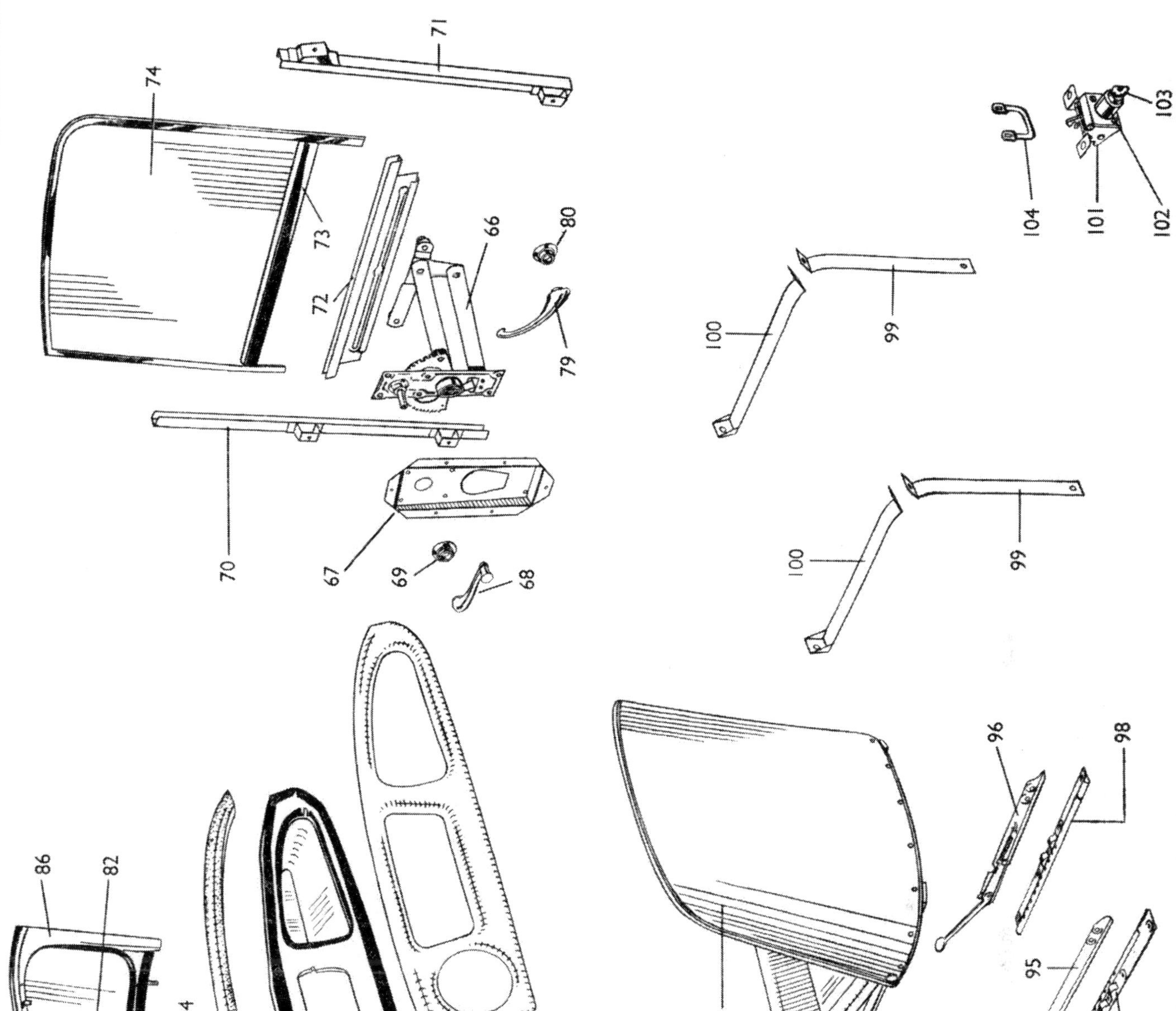

66
67
68
69
70
71
72
73
74
75
76
77
78
79
80
81
82
83
84
85
86
87
88
89
90
91
92
93
94
95
96
97
98
99
100
101
102
103
104

PLATE 29

BODY MOUNTING DETAILS

Part No.	Description	Plate No.	No. per Unit	Remarks
	BODY MOUNTINGS			
12206/12	**Screw,** Set, mounting Body		28	
12206/14	**Screw,** Set, mounting Body		2	
UFN.137/L	**Nut,** on Front Mounting Setscrews		4	
FG.106/X	**Washer,** Spring, under Nuts		4	
400994	**Mounting Rubber**		30	
265809	**Locating Washer,** for Mounting Rubbers		30	
BD.6653	**Packing Pad,** for Body Mounting		As req'd	
BD.2070	**Washer,** Fibre, for Body Mounting		As req'd	
431366	**STEERING COLUMN SUPPORT CLAMP**		1	
UFS.125/7R	**Screw,** Set, securing Clamp		2	
FG.104/X	**Washer,** Spring, on Setscrews		2	
432276	**RUBBER GROMMET IN DASH FOR STEERING COLUMN**		1	
308119	**STAY (LONG) SUPPORTING STEERING COLUMN**		1	R.H. DRIVE CARS ONLY
308120	**STAY (SHORT) SUPPORTING STEERING COLUMN**		1	R.H. DRIVE CARS ONLY
308094	**STAY (LONG) SUPPORTING STEERING COLUMN**		1	L.H. DRIVE CARS ONLY
308095	**STAY (SHORT) SUPPORTING STEERING COLUMN**		1	L.H. DRIVE CARS ONLY
UFB.131/26R	**Bolt,** securing Stay to Clamp		1	
UFN.131/L	**Nut,** on Bolt		1	
FG.105/X	**Washer,** Spring, under Nut		1	
308099	**BLANKING PLATE FOR HEATER APERTURE IN DASH**		1	NOT REQUIRED WHEN HEATER IS FITTED
308098	**Rubber Seal,** for Blanking Plate		1	NOT REQUIRED WHEN HEATER IS FITTED
UFS.125/5R	**Screw,** Set, securing Blanking Plate		4	NOT REQUIRED WHEN HEATER IS FITTED
UFN.125/L	**Nut,** on Setscrew		4	NOT REQUIRED WHEN HEATER IS FITTED
425279	**RUBBER BLANKING PLUG IN DASH FOR REDUNDANT HEATER CONTROL CABLE**		1	NOT REQUIRED WHEN HEATER IS FITTED
81/634556	**LEATHER GAITER FOR GEAR LEVER**		1	NOT REQUIRED WHEN AUTOMATIC TRANSMISSION IS FITTED

Part No.	Description	Plate No.	No. per Unit	Remarks
136150	**HEADLAMP COMPLETE (58260/D)**		2	**R.H. DRIVE CARS ONLY**
9953	**Rim,** (54520073)	30—1	2	
3446	**Dust Excluder,** (552906)	30—2	2	
6518	**Plate,** retaining Light Unit (554872)	30—3	2	
134316/04	**Light Unit,** (553921)	30—4	2	
6057	**Adaptor,** (554602)	30—5	2	
135122/03	**Bulb,** 12v—40/50w (414)	30—6	2	
9668	**Rubber Gasket** (54520131)	30—7	2	
136152	**HEADLAMP, COMPLETE (58560/A)**		2	**L.H. DRIVE CARS EXPORT EUROPE**
7762	**Light Unit,** (556452)		2	
7764	**Adaptor,** complete with Cables (553738)		2	
7763	**Bulb,** 12v—40/45w (410)		2	
	All other items are as for Headlamp 136150			
136199	**HEADLAMP, COMPLETE (58561/A)**		2	**L.H. DRIVE CARS EXPORT FRANCE**
7762	**Light Unit,** (556452)		2	
7764	**Adaptor,** complete with Cables (553738)		2	
8612	**Bulb,** 12v—40/45w (Yellow) (411)		2	
	All other items are as for Headlamp 136150			
136151	**HEADLAMP, COMPLETE (58479/D)**		2	**L.H. DRIVE CARS AND EXPORT U.S.A.**
136151/05	**Rim,** (54520443)		2	
3446	**Dust Excluder,** (552906)		2	
8077	**Plate,** retaining Light Unit (554179)		2	
3454	**Adaptor,** complete with Cables (553724)		2	
8172	**Rubber Gasket,** (554178)		2	
C.20464	**HEADLAMP, COMPLETE (58631)**		2	**R.H. DRIVE CARS EXPORT SWEDEN**
136150/01	**Rim,** (54520073)		2	
3446	**Dust Excluder,** (552906)		2	
6518	**Plate,** retaining Light Unit (554872)		2	
5221	**Screw,** Unit Rim (186137)		6	
6119	**Rim,** seating Unit (554871)		2	
8680	**Light Unit,** (54520148)		2	
7765	**Spring,** retaining Bulb (556442)		2	
7764	**Bulb Holder,** (553738)		2	
8064	**Terminal Sleeve,** (555910)		2	
7763	**Bulb,** (45/40w) (410)		2	
7766	**Gasket,** seating Lamps (554280)		2	
6521	**Trimmer Screw,** (554288)		6	
3434	**Trimmer Spring,** (552813)		6	
9954	**Sleeve,** on Trimmer Screws (554626)		6	
425286/15	**Screw,** securing Headlamps		8	
308032	**SIDELAMP (52426/B)**		2	
307794/01	**Rim,** (516556)	30—8	2	
307794/02	**Screw,** fixing Rim (136231)		2	

Part No.	Description	Plate No.	No. per Unit	Remarks
	SIDELAMP (Cont.)			
307794/03	**Lens,** (516386)	30—9	2	
307794/04	**Gasket,** sealing Lens (136395)	30—10	2	
3455	**Adaptor,** (553780)		2	
989	**Bulb,** 12v—6w (989)	30—11	2	
307794/05	**Seating Rubber,** for Sidelamp (516533)	30—12	2	
425286/20	**Screw,** securing Sidelamp to Body		4	
307795	**FRONT FLASHER LAMP (52337/B)**		2	
307795/01	**Body,** (Rubber) (576105)	30—13	2	
3485	**Rim,** (572734)	30—14	2	
307795/02	**Lens,** (Amber) (576108)	30—15	2	
307795/03	**Bulb Holder Assembly,** (576986)	30—16	2	
C.9126	**Bulb,** 12v—21w (382)	30—17	2	
307796	**FRONT FLASHER LAMP (52345)**		2	FITTED TO EARLIER L.H. DRIVE CARS FOR EXPORT EXCEPT EUROPE
307796/01	**Lens,** (White), (576983)		2	
	All other items are as for Front Flasher Lamp 307795			
425286/20	**Screw,** securing Front Flasher Lamps to Body		4	
308085	**REAR FLASHER LAMP (52459/B)**		2	
305149/02	**Body Assembly** (574234)	30—18	2	
6538	**Lens,** (Amber), (573832)	30—19	2	
83/638066	**Seating Rubber,** for Lens (575640)	30—20	2	
C.9126	**Bulb,** 12v—21w (382)	30—21	2	
C.8460	**Seating Rubber,** for Body (574231)	30—22	2	
308086	**REAR FLASHER LAMP (52460/B)**		2	L.H. DRIVE CARS EXPORT EXCEPT EUROPE
308086/01	**Lens,** (Red), (54570779)		2	
	All other items are as for Rear Flasher Lamp 308085			
426934/15	**Screw,** securing Rear Flasher Lamps to Body		6	
308087	**STOP/TAIL LAMP (53818/B)**		2	
5312	**Body Assembly,** (574224)	30—23	2	
305150/04	**Lens,** (574226)	30—24	2	
83/638066	**Seating Rubber,** for Lens (575640)	30—25	2	
C.9125	**Bulb,** 12v—21/6w (380)	30—26	2	
C.8460	**Seating Rubber,** for Body (574231)	30—27	2	
426934/15	**Screw,** securing Stop/Tail Lamps		6	
308088	**REFLEX REFLECTOR (575883)**	30—28	2	

Part No.	Description	Plate No.	No. per Unit	Remarks
136168	**FINISHER FOR STOP/TAIL AND REAR FLASHER LAMPS**	30—29	2	
431462/38	**Screw,** securing Finisher		4	
431462/40	**Screw,** securing Finisher		4	
308061	**SEALING RUBBER UNDER FINISHER**	30—30	2	
305148	**REAR NUMBER PLATE ILLUMINATION LAMP (052410/F)**		1	**NOT FITTED TO CARS WITH FULL LENGTH REAR BUMPER**
305148/01	**Body,** (572603)	30—31	1	
305148/02	**Window,** (572596)	30—32	1	
305148/03	**Gasket,** seating Window (572595)	30—33	1	
305148/04	**Base Assembly,** (572619)	30—34	1	
989	**Bulb,** 12v—6w (989)	30—35	1	
305152	**REAR NUMBER PLATE ILLUMINATION LAMP (53278/B)**		1	**FITTED TO CARS WITH FULL LENGTH REAR BUMPER**
305152/01	**Cover,** (574046)		1	
305152/03	**Seating Rubber,** for Cover (574044)		1	
989	**Bulb,** 12v 6w (989)		1	
NB.119/5R	**Bolt,** securing Number Plate Illumination Lamp		2	
NN.119/L	**Nut,** on Bolts		2	
AG.102/X	**Washer,** Spring, under Nuts		2	
135093	**HORN (LOW NOTE) (WT.618/69046/F)**	31—36	1	**1st CONDITION**
135094	**HORN (HIGH NOTE) (WT.618/69047/F)**		1	**1st CONDITION**
136577	**HORN (LOW NOTE) (9H/54068009)**		1	**2nd CONDITION**
136578	**HORN (HIGH NOTE) (9H/54068008)**		1	**2nd CONDITION**
308162	**BRACKET ASSEMBLY MOUNTING HORNS**	31—37	2	
UFS.125/5R	**Screw,** Set, securing Brackets to Chassis		4	**1st CONDITION**
UFN.125/L	**Nut,** on Setscrews		4	**1st CONDITION**
FG.104/X	**Washer,** Spring, under Nuts		4	**1st CONDITION**
UFB.125/14R	**Bolt,** securing Brackets to Chassis		4	**2nd CONDITION**
432079	**Distance Piece,** on Bolts		4	**2nd CONDITION**
432483	**Grommet,** for Horns		2	**2nd CONDITION**
UFS.125/6R	**Screw,** Set, securing Horns to Brackets		4	
UFN.125/L	**Nut,** on Setscrews		4	
FG.104/X	**Washer,** Spring, under Nuts		4	

Part No.	Description	Plate No.	No. per Unit	Remarks
308036	**BATTERY (BT.9A) (4016417)**	31—38	1	
307935	**Angle Assembly,** securing Battery	31—39	1	
431431	**Rod,** securing Battery	31—40	2	
88602	**Wing Nut,** on Rods		2	
FW.104/T	**Washer,** Plain, under Wing Nuts		2	
307994	**PETROL TANK ELEMENT UNIT (FT. 5301/05)**	31—41	1	
307994/01	**Cork Washer,** seating Element Unit	31—42	1	
426934/22	**Screw,** securing Element Unit to Petrol Tank		6	
431723	**SWITCH, OPERATING STOP LIGHTS (398003)**	31—44	1	
431438	**Gasket,** for Switch (378707)		1	
431429	**DIPPER SWITCH (31800/B—21 SA.)**	31—45	1	
425287/20	**Screw,** securing Dipper Switch		2	
431430	**SOLENOID FOR STARTER MOTOR (76464/D—2 ST.)**	31—46	1	NOT REQUIRED ON CARS FITTED WITH AUTOMATIC TRANSMISSION
308723	**SOLENOID FOR STARTER MOTOR (76469—2 ST.)**		1	REQUIRED ONLY ON CARS FITTED WITH AUTOMATIC TRANSMISSION
425286/17	**Screw,** securing Solenoid		2	
423199/2	**Washer,** Plain, on Screws		2	
308035	**CONTROL BOX (37299/D—RB.310)**	31—47	1	
90018/08	**Screw,** securing Control Box		3	
C.16729	**FLASHER UNIT (35020/A)**	31—48	1	
90018/06	**Screw,** securing Flasher Unit		1	
308097	**FUSE UNIT (54038033)**	31—49	1	
425256/17	**Screw,** securing Fuse Unit		1	

Part No.	Description	Plate No.	No. per Unit	Remarks
C.9688/1	**PETROL PUMP (AUA.52)**		1	**1st CONDITION**
307855/01	**Body,** (AUA.4644)	31—50	1	
307855/02	**Lid,** for Body (top and bottom) (AUA.4645)	31—51	2	
1833	**Gasket,** under Lids (AUA.4646)	31—52	2	
307855/04	**Screw,** securing Lids	31—53	12	
C.723	**Washer,** Shakeproof on Screws (AUA.4643)	31—54	12	
307855/05	**Valve Cage,** (AUA.4607)	31—55	1	
307855/06	**Valve Disc,** (AUA.4611)	31—56	2	
307855/07	**Clip,** for Valve Cage (AUA.4609)	31—57	1	
5846	**Filter,** (AUA.4647)	31—58	1	
C.11488	**Banjo Bolt,** (AUC.2698)	31—59	2	
C.11489	**Washer,** Fibre, on Banjo Bolts (AUC.2141)	31—60	4	
1238	**Diaphragm Assembly** (AUA.6010)	31—61	1	
10201	**Armature Roller,** (AUA.1433)	31—62	11	
1272	**Spring,** for Armature (AUA.1785)	31—63	1	
307855/12	**Coil Housing Assembly** (AUA.6003)	31—64	1	
7124	**Screw,** securing Coil Housing (AUA.1453)	31—65	6	
307855/14	**Contact Breaker Spindle** (AUA.1435)	31—66	1	
6046	**Rocker and Blade Assembly,** (Double Contact) (AUA.6021)	31—67	1	
307855/17	**Screw,** securing Blade (AUA.847)	31—69	1	
307855/18	**Washer,** Spring, on Screw (AUA.868)	31—70	1	
307855/19	**Pedestal Assembly,** (AUA.6034)	31—71	1	
307855/20	**Screw,** securing Pedestal (AUA.1459)	31—72	2	
307855/21	**Washer,** Spring, on Screws (AUA.1863)		2	
7664	**Condenser,** (AUA.5060)	31—73	1	
7665	**Clip,** holding Condenser (AUA.5059)	31—74	1	
1145	**Terminal Screw,** (AUA.1468)	31—75	1	
307855/23	**Nut,** on Terminal Screw (AUA.1661)	31—76	1	
307855/24	**Washer,** on end of Cable for Terminal Screw (AUA.1662)	31—77	1	
307855/21	**Washer,** Spring, on Terminal Screw (AUA.1863)	31—78	1	
7666	**End Cover,** (AUA.5078)	31—79	1	
307855/31	**Dust Excluder,** for End Cover (AUA.504)	31—80	1	
307855/26	**Terminal Nut,** (AUA.878)	31—81	2	
307855/32	**Washer,** Fibre, under Terminal Nut (AUA.505)	31—82	1	
C.723	**Washer,** Shakeproof, under Terminal Nut (AUA.4643)	31—83	1	
C.18336	**PETROL PUMP (AUA.152)**		1	**2nd CONDITION**
1833	**Gasket,** under Lids (AUA.4646)		2	
8754	**Diaphragm Assembly,** (AUB.6004)		1	
9320	**Spring,** on Diaphragm (AUB.521)		1	
5846	**Filter,** (AUA.4647)		1	
6046	**Rocker and Blade Assembly** (AUB.6021)		1	
7664	**Condenser,** (AUA.5060)		1	
7665	**Clip,** holding Condenser (AUA.5059)		1	
7666	**End Cover,** (AUA.5078)		1	
	All other items are as for Petrol Pump C.9688/1			
C.20576	**PETROL PUMP (AUB.152)**		1	**3rd CONDITION**
9098	**Diaphragm Assembly** (AUB.6040)		1	
9099	**Gasket,** between Diaphragm and Valve Housing (AUB.593)		2	
9100	**Rocker and Blade Assembly** (AUB.6022)		1	
1145	**Terminal Screw,** (AUA.1468)		1	
9321	**End Cover,** (AUB.626)		1	
	All other items are as for Petrol Pump C.9688/1			

ELECTRICAL EQUIPMENT

Part No.	Description	Plate No.	No. per Unit	Remarks
136251	**WINDSCREEN WIPER MOTOR ASSEMBLY (75371/E)**	32—84	1	
136251/01	**Crosshead and Rack (743223)**	32—85	1	
308233	**OUTER CASING, FROM MOTOR TO WHEELBOX (54711646)**	32—86	1	
308234	**OUTER CASING, WHEELBOX TO WHEELBOX (54711647)**	32—87	1	
C.5450	**OUTER CASING FOR WHEELBOX EXTENSION (740746)**	32—88	1	
423393	**GROMMET IN DASH FOR CASING**	32—89	1	
308252	**WHEELBOX ASSEMBLY (726930)**	32—90	2	
308251/01	**ARM, FOR WINDSCREEN WIPER (54711565)**	32—91	2	R.H. DRIVE CARS ONLY
308250/01	**ARM, FOR WINDSCREEN WIPER (54711566)**	32—91	2	L.H. DRIVE CARS ONLY
51/638001	**BLADE, FOR WINDSCREEN WIPER ARMS (738724)**	32—92	2	
308075	**SPEEDOMETER (SN.6321/02) (CALIBRATED IN MILES)**	32—93	1	
308082	**SPEEDOMETER (SN. 6321/03) (CALIBRATED IN KILOS)**		1	
987	**Bulb,** for Speedometer illumination, Ignition and Main Beam Warning Lights (12v—2·2w) (987)	32—94	3	
308083	**SPEEDOMETER FLEXIBLE CABLE COMPLETE (DF.1104/00/55)**	32—95	1	R.H. DRIVE CARS ONLY NOT REQUIRED WHEN AUTOMATIC TRANSMISSION IS FITTED
308083/01	**Inner Cable**	32—96	1	
308083/02	**Outer Cable**	32—97	1	

Part No.	Description	Plate No.	No. per Unit	Remarks
308005	**SPEEDOMETER FLEXIBLE CABLE COMPLETE (DF.1104/00/63)**		1	L.H. DRIVE CARS ONLY NOT REQUIRED WHEN AUTOMATIC TRANSMISSION IS FITTED
308005/01	**Inner Cable**		1	
308005/02	**Outer Cable**		1	
308327	**SPEEDOMETER FLEXIBLE CABLE COMPLETE (DF.1302/01/60)**		1	REQUIRED ON ALL CARS WHEN AUTOMATIC TRANSMISSION IS FITTED
308327/01	**Inner Cable** (DI. 1314/00/60)		1	
308327/02	**Outer Cable** (DO 1314/00/60)		1	
424759	**Clip,** securing Speedometer Cable to Floor		1	NOT REQUIRED WHEN AUTOMATIC TRANSMISSION IS FITTED
NS.419/6H	**Screw,** securing Clip		1	
UFN.119/L	**Nut,** on Screw		1	
AG.102/X	**Washer,** Spring, under Nut		1	
431938	**Clip,** securing Speedometer Cable to Gearbox Cover		1	
432292	**Grommet,** in Floor for Speedometer Cable	32—98	1	
UCS.125/3R	**Screw,** securing Speedometer Cable to Transmission Unit		1	REQUIRED ON ALL CARS WHEN AUTOMATIC TRANSMISSION IS FITTED
FG.104/X	**Washer,** Spring, on Setscrew		1	
432451	**Clip,** securing Speedometer Cable to Heater Mounting Bolt		1	R.H. DRIVE CARS ONLY REQUIRED WHEN AUTOMATIC TRANSMISSION IS FITTED
432909	**Clip,** securing Speedometer Cable under Front Floor		1	
423187/24	**Screw,** securing Clip		1	
308075	**REV COUNTER (RN.2410/01)**	32—99	1	
987	**Bulb,** illuminating Rev Counter and Flasher Warning Lights (987)	32—100	3	
308084	**REV COUNTER FLEXIBLE CABLE (DF.1104/00/40)**	32—101	1	R.H. DRIVE CARS ONLY NOT REQUIRED WHEN AUTOMATIC TRANSMISSION IS FITTED
308084/01	**Inner Cable**	32—102	1	
308084/02	**Outer Cable**	32—103	1	
309145	**REV COUNTER FLEXIBLE CABLE (DF.1104/00/48)**		1	R.H. DRIVE CARS ONLY REQUIRED WHEN AUTOMATIC TRANSMISSION IS FITTED
309145/01	**Inner Cable**		1	
309145/02	**Outer Cable**		1	
308006	**REV COUNTER FLEXIBLE CABLE (DF.1104/00/30)**		1	L.H. DRIVE CARS ONLY
308006/01	**Inner Cable**		1	
308006/02	**Outer Cable**		1	
431536	**Clip,** securing Cable to Facia Stay		1	
432292	**Grommet,** in Dash, for Cable	32—104	1	

Part No.	Description	Plate No.	No. per Unit	Remarks
308077	**PETROL GAUGE (FG.2331/06)**	32—105	1	
987	**Bulb,** illuminating Petrol Gauge (12v—2·2w) (987)	32—106	1	
308079	**WATER TEMPERATURE GAUGE (TC.4304/01)**	32—107	1	
987	**Bulb,** illuminating Temperature Gauge (12v—2·2w) (987)	32—108	1	
308080	**WATER TEMPERATURE TRANSMITTER (TT.3800/00)**	32—109	1	
308884	**OIL PRESSURE GAUGE (PL.2303/04)**	32—110	1	
987	**Bulb,** illuminating Oil Pressure Gauge (12v—2·2w) (987)	32—111	1	
308621	**PIPE ASSEMBLY, FOR OIL PRESSURE GAUGE**		1	
431515	**Washer,** for Pipe Union		1	
432280	**Grommet,** in Dash, for Pipe		1	
425456	**Clip,** securing Pipe to Dash		1	
426933/34	**Screw,** securing Clip		1	
AG.102/X	**Washer,** Spring, on Screw		1	
308096	**AMMETER (CZU.60—36263/A)**	32—112	1	
987	**Bulb,** illuminating Ammeter (12v—2·2w) (987)	32—113	1	
431424	**LIGHTING SWITCH (57 SA.—31788)**	32—114	1	1st CONDITION
308824	**LIGHTING SWITCH (57 SA.—34351)**		1	2nd CONDITION
431531	**Designation Plate,** "LIGHTS"	32—115	1	
431426	**SWITCH PANEL LIGHTS (65 SA—31851)**	32—116	1	1st CONDITION
308825	**SWITCH, PANEL LIGHTS (65 SA—34352/B)**		1	2nd CONDITION
431530	**Designation Plate,** "PANEL"	32—117	1	
431425	**STARTER AND IGNITION SWITCH (47 SA.—31799/D)**	32—118	1	

Part No.	Description	Plate No.	No. per Unit	Remarks
424926	**IGNITION BARREL**	32—119	1	
424927	**Key,** for Ignition Barrel	32—120	2	
431427	**SWITCH, WINDSCREEN WIPER (57 SA—31780)**	32—121	1	1st CONDITION
308826	**SWITCH, WINDSCREEN WIPER- (57 SA—34350/E)**		1	2nd CONDITION
431532	**Designation Plate,** "WIPER"	32—122	1	
308103	**INSTRUMENT PANEL**	32—123	1	
87105/04	**Screw,** securing Instrument Panel		9	
136290	**MAIN HARNESS (54943950)**		1	
308354	**PANEL HARNESS (54944244)**		1	
308353	**WINDSCREEN WIPER HARNESS (54944262)**		1	
	EARTH CABLES			
431884	**Earth Cable,** for Switch Panel (54944260)		3	
431885	**Earth Cable,** for Tail (54944261)		4	
431878	**Earth Cable,** for Battery (54943957)		1	
431879	**Earth Cable,** for Engine (54943958)		1	
308542	**Earth Cable,** for Gearbox (54945917)		1	
431884	**Earth Cable,** for Switch Panel (54944260)		1	L.H. DRIVE CARS ONLY
	CABLES			
431880	**Cable,** from Battery to Solenoid (54943960)		1	R.H. DRIVE CARS ONLY
431886	**Cable,** from Battery to Solenoid (54943959)		1	L.H. DRIVE CARS ONLY
308356	**Cable,** Steering Column Extension (54943954)		1	L.H. DRIVE CARS ONLY
431883	**Cable,** for Instrument Panel Light (54944243)		2	
308351	**Cable,** Extension for Horns (54943953)		1	1st CONDITION
309015	**Cable,** Extension for Horns (54948689)		1	2nd CONDITION
308352	**Cable,** for Stop Light Switch (54943955)		1	
431881	**Cable,** for Front Flasher Lamp L.H. (54944240)		1	
431882	**Cable,** for Front Flasher Lamp R.H. (54944241)		1	
308349	**Cable,** for Stop/Tail Lamp L.H. (54943951)		1	
308350	**Cable,** for Stop/Tail Lamp R.H. (54943952)		1	
308355	**Cable,** from junction to Petrol Unit (54944259)		1	
431877	**Cable,** from Solenoid to Starter (54943956)		1	

Part No.	Description	Plate No.	No. per Unit	Remarks
	GROMMETS			
432293	**Grommet,** in Valance for Headlamp and Sidelamp Cables		2	
432424	**Grommet,** in Front Wing for Sidelamp Cables		2	
432293	**Grommet,** for Tail Lamp Cables		2	
432293	**Grommet,** in Bulkhead for Harness		1	
	CABLE CLIPS			
408619	**Clip,** for Front Harness		3	**R.H. DRIVE CARS ONLY**
411480	**Clip,** for Front Harness		3	**L.H. DRIVE CARS ONLY**
425286/15	**Screw,** securing Clips		3	
431960	**Clip,** for Dynamo Cable		1	
UCS.125/4R	**Screw,** Set, securing Clip		1	
FG.104/X	**Washer,** Spring, on Setscrew		1	
431034	**Clip,** securing Battery Cable		1	

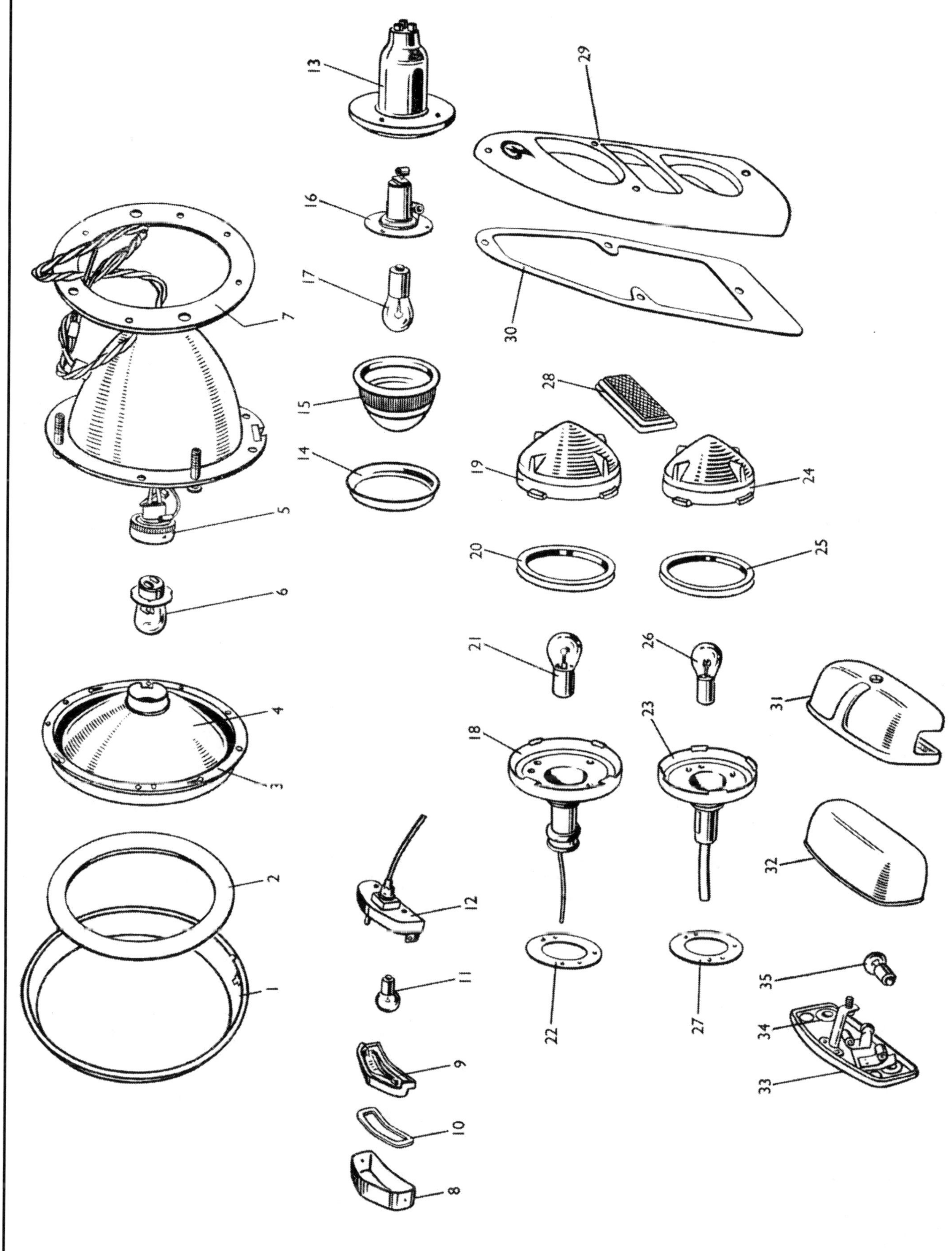
1
2
3
4
5
6
7
8
9
10
11
12
13
14
15
16
17
18
19
20
21
22
23
24
25
26
27
28
29
30
31
32
33
34
35

PLATE 30

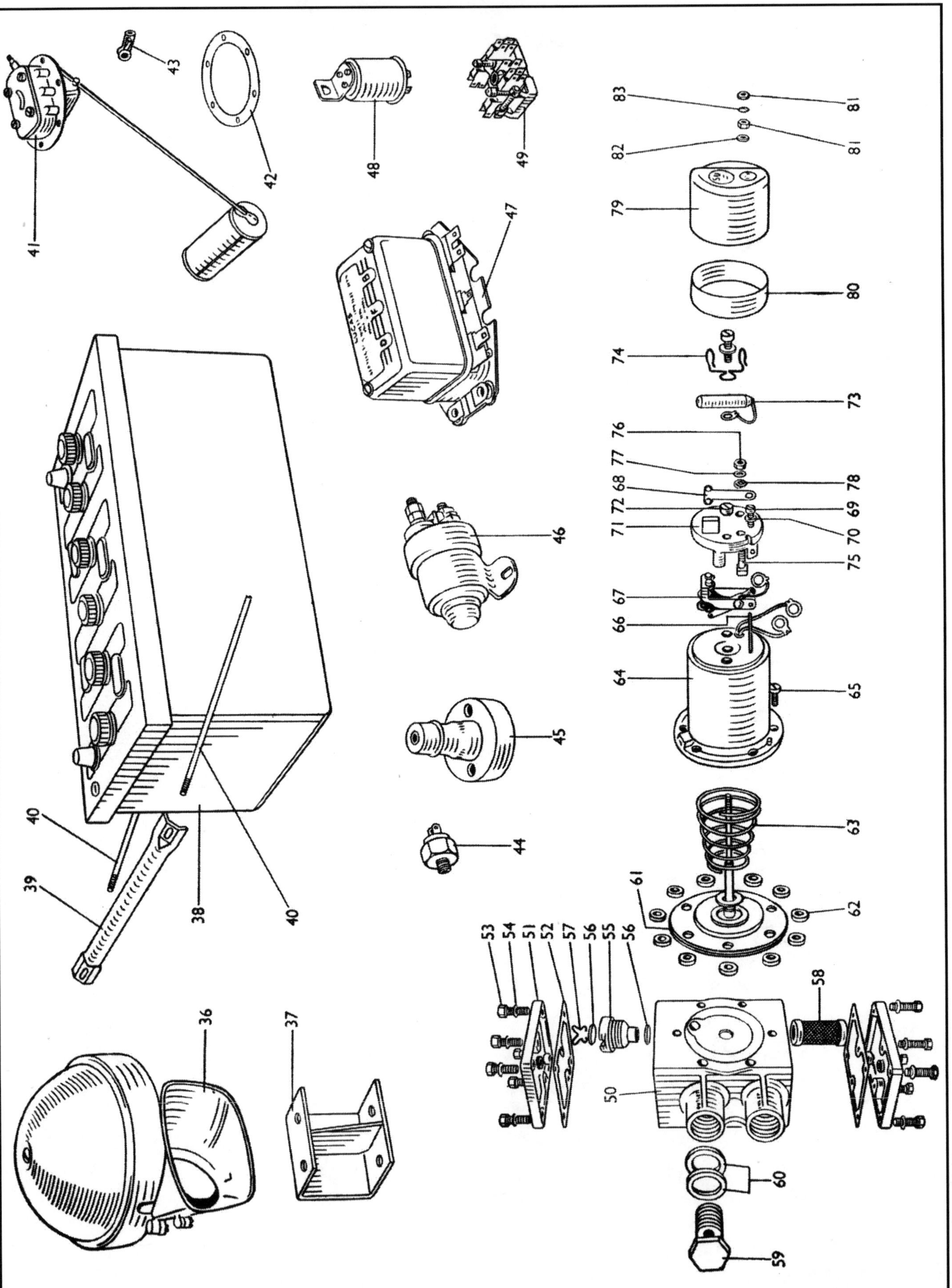
36
37
38
39
40
40
41
42
43
44
45
46
47
48
49
50
51
52
53
54
55
56
56
57
58
59
60
61
62
63
64
65
66
67
68
69
70
71
72
73
74
75
76
77
78
79
80
81
81
82
83

PLATE 31

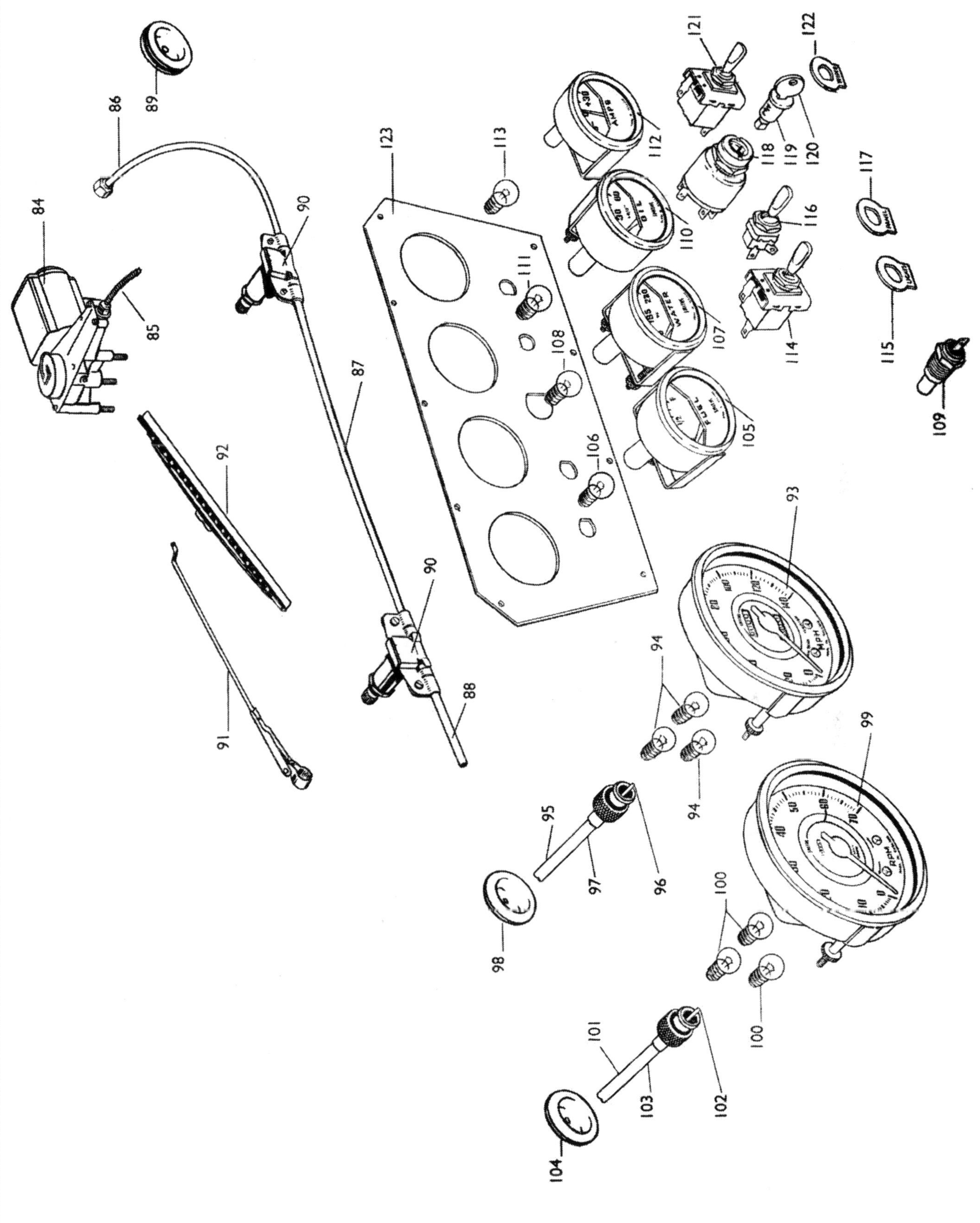
84
85
86
87
88
89
90
90
91
92
93
94
94
95
96
97
98
99
100
100
101
102
103
104
105
106
107
108
109
110
111
112
113
114
115
116
117
118
119
120
121
122
123

PLATE 32

TOOL KIT

Part No.	Description	Plate No.	No. per Unit	Remarks
424754	**Double-ended Spanner** (U.N. No. 8 x 10)		1	
424755	**Double-ended Spanner** ($\frac{7}{16}$" × $\frac{1}{2}$" A/F.)		1	
303793	**Double-ended Spanner** ($\frac{9}{16}$" × $\frac{5}{8}$" A/F)		1	
303794	**Double-ended Spanner** ($\frac{11}{16}$" × $\frac{3}{4}$" A/F)		1	
307553	**Box Spanner,** for Sparking Plugs		1	
303789	**Tommy Bar,** for Box Spanner		1	
303102	**Wheel Brace**		1	} FOR CARS FITTED WITH
425706	**Hub Cap Tool**		1	} DISC TYPE ROAD WHEELS
426071	**Screwdriver and Points Gauge**		1	
418259	**Pliers**		1	
308011	**Screwdriver**		1	
431490	**Tappet Gauge**		1	
307984	**Lifting Jack**		1	
307985	**Lifting Jack Spanner**		1	
424752	**Bleeder Tube,** complete with container		1	
134311	**Tool Roll**		1	
134312	**Tool Bag**		1	
308089	**Hammer** (Lead-faced)		1	FOR CARS FITTED WITH WIRE SPOKE ROAD WHEELS

Part No.	Description	Plate No.	No. per Unit	Remarks
	HEATER AND DEMISTER			
308151	**WATER INLET PIPE FOR HEATER**	33—1	1	
136212	**WATER OUTLET PIPE FOR HEATER**	33—2	1	
422255	**Elbow,** connecting Water Pipes to Heater	33—3	2	
422522/02	**Clip,** securing Elbows		4	
431946	**Clip,** securing Water Outlet Pipe to Induction Manifold	33—4	1	
422545/13	**Screw,** securing Clip		1	
422543/2	**Nut,** on Screw		1	
AG.102/X	**Washer,** Plain, on Screw		1	
136209	**HEATER UNIT ASSEMBLY (FHR.3454/01)**	33—5	1	
308072	**Seal,** between Heater Unit Base and Dash	33—6	1	
UFS.125/6R	**Screw,** Set, securing Heater Unit to Dash		4	
FW.104/T	**Washer,** Plain, on Setscrews		4	
308135	**HEATER CONTROL UNIT, COMPLETE WITH BLOWER SWITCH (FHC.1230/11)**	33—7	1	
431610	**Knob,** on Heater Control Unit Lever (CH.85216/01)	33—8	1	
431734	**Screw,** securing Knob (30-257-104-11)		1	
308109	**Indicator Strip,** on Heater Control Unit	33—9	1	
422544/9	**Screw,** securing Heater Control Unit		2	
FG.103/X	**Washer,** Spring, on Screws		2	
431611	**HEATER CONTROL CABLE (FHC.8401/96)**	33—10	1	
431612	**Trunnion,** for Control Cable (CH.31701/01)	33—11	1	
432279	**Grommet,** in Dash for Control Cable		1	
308186	**AIR CONTROL CABLE ASSEMBLY (FHC.6151/10)**	33—12	1	
308186/01	**Operating Knob,** on Control Cable (CH.85129/1)	33—13	1	
431613	**Trunnion,** on Air Control Cable (CH.31502)	33—14	1	
431585/1	**BRACKET, FOR AIR CONTROL CABLE**		1	
423193/12	**Woodscrew,** securing Bracket to Dash		2	
308163	**DEMISTER TUBE (22″ LONG) (FHD.7115/PAB/88)**	33—15	1	

OPTIONAL EXTRAS

Part No.	Description	Plate No.	No. per Unit	Remarks
308164	**DEMISTER TUBE (27″ LONG) (FHD.7115/PAB/108)**	33—16	1	
	WIRE SPOKE ROAD WHEELS			
308090	**R.H. REAR HUB FOR WIRE SPOKE ROAD WHEEL**		1	**Replacing "Rear Axle" Parts:—** 307996, 430649 425658/02, 430450 431685
308091	**L.H. REAR HUB FOR WIRE SPOKE ROAD WHEEL**	34—1	1	
431511	**Bolt,** securing Brake Disc to Hubs		8	
136244	**R.H. FRONT SUSPENSION UNIT ASSEMBLY**		1	**Replacing "Front Suspension" Parts:—** 136194 136195 308174 431385 431386 430241/06 430650
308092	**R.H. Front Hub**		1	
431512	**Bolt,** securing Brake Discs to Hub		4	
UFN.137/L	**Nut,** on Bolts		4	
136245	**L.H. FRONT SUSPENSION UNIT ASSEMBLY**		1	
308093	**L.H. Front Hub**	34—2	1	
431515	**Bolt,** securing Brake Discs to Hub		4	
UFN.137/L	**Nut,** on Bolts		4	
136201	**WIRE SPOKE ROAD WHEEL (TYPE 42W)**		5	**Replacing "Road Wheels" Parts:—** 136198 136196
136201/01	**Long Spoke**		100	
136201/02	**Short Spoke**		200	
136201/03	**Nipple**		300	
308104	**R.H. HUB CAP**		2	
308105	**L.H. HUB CAP**	34—3	2	
81/634295	**COVER, OVER SPARE WHEEL CLAMP**	34—4	1	**Replacing "BODY DETAILS" Parts:—** 87625
87103/03	**Woodscrew,** securing Cover to Floor		4	
87627	**Bolt,** for Spare Wheel Clamp	34—5	1	

Part No.	Description	Plate No.	No. per Unit	Remarks
	ADJUSTABLE STEERING COLUMN			
136227	**STEERING COLUMN CONTROL ASSEMBLY** (33556)		1	
136227/01	**Top Cover Assembly** (54300427)		1	
136227/02	**Stator Plate Assembly** (54300443)		1	
136227/03	**Cable** (54944643)		1	
136227/01	**TOP COVER ASSEMBLY** (54300427)		1	
135246/02	**Horn Push** (322175)		1	
133055/02	**Ring**, retaining Horn Push (322176)		1	
308292	**STATOR TUBE ASSEMBLY** (54300449)		1	
136184	**STEERING WHEEL ASSEMBLY (ALTERNATIVE TO 431842)**	34—6	1	**Replacing "Steering Unit and Controls" Parts:—** 135932, 135932/01, 135932/02, 135932/03, 135932/05, 135932/06, 135990, 427234, 427235, 405066, 136057, 136058, 136057/01, 136058/01, 135217/04, 136057/02, 136058/03
431842	**STEERING WHEEL ASSEMBLY (LEATHER COVERED) (ALTERNATIVE TO 136184)**			
136184/01	**Locking Sleeve,** for Steering Wheel	34—7	1	
131752/02	**Volute Spring**	34—8	1	
131752/03	**Washer,** for Top Dust Cover	34—9	1	
131752/04	**Washer,** for Bottom Dust Cover	34—10	1	
279995	**Circlip,** securing Steering Wheel to Inner Column	34—11	1	
424806	**Grubscrew,** securing Steering Column Control		2	
136185	**STEERING UNIT ASSEMBLY (Z.1417) (R.H. DRIVE CARS ONLY)**		1	
136185/01	**Outer Column and Steering Box** (37873)		1	
133502/02	**Bush,** Felt, for Outer Column (8533)		1	
136185/02	**Inner Column and Worm** (37826)		1	
136186	**STEERING UNIT ASSEMBLY (Z.1418) (L.H. DRIVE CARS ONLY)**		1	
136186/01	**Outer Column and Steering Box** (37877)		1	
133502/02	**Bush,** Felt, for Outer Column (8533)		1	
136186/02	**Inner Column and Worm** (37828)		1	
	WHEEL DISCS (For Disc Type Road Wheels)			
136269	**WHEEL DISC ASSEMBLY ("D" TYPE)**	34—12	4	**Replacing "Road Wheel' Parts:—** 136196
136269/01	**Wheel Disc**	34—13	4	
136269/02	**Inner Adaptor Assembly**	34—14	4	
136269/03	**Outer Adaptor**	34—15	4	
136269/04	**Cap Assembly**	34—16	4	
136269/05	**Cap Rubber**	34—17	8	

OPTIONAL EXTRAS

Part No.	Description	Plate No.	No. per Unit	Remarks
	'SCREENJET' WINDSCREEN WASHER			
C.17004	**WINDSCREEN WASHER UNIT (54071196)**	34—18	1	**STANDARD EQUIPMENT ON LATER R.H. DRIVE CARS. OPTIONAL EXTRA ON ALL L.H. DRIVE CARS**
8408	**Motor,** complete with Coupling (54711928)	34—19	1	
8035	**Pump Assembly** (295194)	34—20	1	
8041	**Bracket,** holding Reservoir (295061)	34—21	1	
8040	**Glass Reservoir** (295058)	34—22	1	
8036	**Moulded Cover,** for Reservoir (295138)	34—23	1	
8037	**Gasket,** Rubber, between Cover and Reservoir (295042)		1	
8039	**Union,** securing Motor to Pump (295045)		1	
8038	**Rubber Filler Cap,** in Cover (295043)	34—24	1	
308207	**Bracket,** mounting Windscreen Washer Unit Bracket to Valance	34—25	1	
422545/12	**Screw,** Set, securing Bracket to Valance		4	
NN.119/L	**Nut,** on Setscrews		4	
AW.102/T	**Washer,** Plain, on Setscrews		4	
AG.102/X	**Washer,** Spring, under Nuts		4	
UCS.313/3H	**Screw,** securing Reservoir Bracket to Mounting Bracket		3	
UCN.113/L	**Nut,** on Screws		3	
430241/01	**Washer,** Spring, under Nuts		3	
8408	**MOTOR, COMPLETE WITH COUPLING (54711928)**	34—19	1	
8042	**Contact Assembly** (295251)		1	
8043	**Brush Set** (295198)		1	
8044	**Armature** (295023)		1	
431626	**TUBING FOR WINDSCREEN WASHER (61″ LONG) (295070)**	34—26	1	
431625	**JET ASSEMBLY (TWIN) (295110)**	34—27	1	
431624	**SWITCH OPERATING WINDSCREEN WASHER (31513/A)**	34—28	1	
431585/03	**Bracket,** supporting Switch		1	
423193/12	**Woodscrew,** securing Bracket to Facia		2	
308709	**HARNESS FOR WINDSCREEN WASHER (54946520)**		1	**R.H. DRIVE CARS ONLY**
308423	**HARNESS FOR WINDSCREEN WASHER (54945455)**		1	**L.H. DRIVE CARS ONLY**
433100	**EARTH LEAD FOR WINDSCREEN WASHER SWITCH**		1	**L.H. DRIVE CARS ONLY**

Part No.	Description	Plate No.	No. per Unit	Remarks
	EXHAUST TAIL PIPE EXTENSION			
308283	**EXTENSION, CHROMIUM PLATED, FOR EXHAUST TAIL PIPE**	34—29	2	
423176/18	**Clip,** securing Extension to Tail Pipe	34—30	2	
	FAN COWL			
136274	**FAN COWL ASSEMBLY**	34—31	1	FOR COUNTRIES WITH HIGH AMBIENT TEMPERATURE
	CIGAR LIGHTER			
309243	**CIGAR LIGHTER (L.15082—12v)**	34—32	1	
C.18638/2	**Element** (LHA.30882—12v)		1	
C.18638/4	**Handle** (LHA.30578—1)		1	
C.18638/5	**Chrome Bezel** (LH.30843—1)		1	
C.18638/6	**Clamping Shell** (LU.30669—3)		1	
C.18638/1	**Switch Base Assembly** (LBA.31075—12v)		1	
C.23606	**EARTH LEAD FOR CIGAR LIGHTER**		1	
	PETROL TANK RESERVE UNIT			
307932	**PETROL TANK RESERVE UNIT (78041)**	34—33	1	Replacing "Fuel System" Parts:— 307987
431634	**Pipe,** from Flexible Pipe to Reserve Unit	34—34	1	
431637	**Union Nut,** securing Pipe to Unit	34—35	1	
308209	**SWITCH OPERATING PETROL RESERVE UNIT (56 SA.—31696)**	34—36	1	
C.15787	**Bulb,** for Switch Warning Light	34—37	1	
431585/2	**Bracket,** supporting Petrol Reserve Switch		1	
423193/12	**Woodscrew,** securing Bracket to Facia		2	
308711	**CABLE FROM SWITCH TO JUNCTION (54946522)**		1	
308710	**EARTH LEAD FOR SWITCH (54946521)**		1	

OPTIONAL EXTRAS

Part No.	Description	Plate No.	No. per Unit	Remarks
	LEATHER COVERED STEERING WHEEL			
431841	**STEERING WHEEL (LEATHER COVERED)**		1	**FOR USE WITH FIXED STEERING COLUMN**
431842	**STEERING WHEEL (LEATHER COVERED)**		1	**FOR USE WITH ADJUSTABLE STEERING COLUMN**
	WHITE WALL TYRES			
431735	**WHITE WALL TYRE (5·90 x 15)**		5	Replacing "Road Wheels" Part:— 430575/1
C.22148	**WHITE WALL TYRE (SLIMLINE) (5·90 x 15)**		5	
	TONNEAU COVER			
81/642930	**TONNEAU COVER ASSEMBLY**		1	**R.H. DRIVE CARS ONLY**
82/642930	**TONNEAU COVER ASSEMBLY**		1	**L.H. DRIVE CARS ONLY**
80480	**Button Fastener**		27	
80469	**Nut,** securing Button Fasteners		27	
80468	**Stud,** receiving Button Fasteners		8	Fitted to Scuttle
88628	**Nut,** securing Studs		8	
	SAFETY BELTS			
9765	**SEAT BELT (LAP/DIAGONAL FITTING) COMPLETE WITH ATTACHMENT BOLTS AND EYE BOLTS**		2	
BD.25479	**'D' MOTIF FOR SEAT BELT BUCKLE**		2	
UCN.111/K	**Nut,** securing 'D' Motif to Buckles		2	
C.720	**Washer,** Shakeproof, under Nuts		2	

Part No.	Description	Plate No.	No. per Unit	Remarks
	HARD TOP COVER			
81/643145	**HARD TOP COVER ASSEMBLY (TRIMMED)**	35—1	1	
81/633070	**Backlight** (Perspex)	35—2	1	FITTED TO EARLY CARS
81/633347	**Backlight** (Toughened Glass)		1	FITTED TO LATER CARS
81/633143	**Glazing Rubber,** for Backlight	35—3	1	
81/633327	**Filler,** Rubber, for Backlight	35—4	1	
81/643180	**Toggle Fastener Assembly**	35—5	3	
87524/05	**Screw,** securing Toggle Fastener		6	
81/633400	**Seating Rubber,** between Body and Hard Top (118" Long)	35—6	1	
423192/11	**Rivet,** securing Seating Rubber to Hard Top		30	
470011	**Finisher,** for Seating Rubber	35—7	1	
81/633333	**Finisher,** at L.H. end of Seating Rubber		1	
81/633334	**Finisher,** at R.H. end of Seating Rubber		1	
90022/03	**Screw,** securing Finishers		4	
BD.23474	**Seating Rubber,** at Front of Hard Top		1	
81/633331	**Weather Strip,** between Hard Top and Windscreen		1	
370152	**Hook,** securing Hard Top to Body	35—8	7	
88527	**Hand Nut,** (Special), securing Hooks	35—9	7	
F.204/X	**Washer,** D.C. Spring		7	
81/633249	**DRAUGHT EXCLUDER AT TOP OF DOOR APERTURE**		2	
	BADGE BAR			
308493	**BADGE BAR**	35—10	1	
432061	**Rubber Mounting,** for Badge Bar	35—11	4	
UFB.137/7R	**Bolt,** securing Badge Bar to Body	35—12	2	
FW.106/T	**Washer,** Plain, on Bolts	35—13	2	
	FOG AND LONG RANGE DRIVING LAMPS			
308452	**FOG LAMP ASSEMBLY (55213—5 FT.)**	35—14	1	
308452/01	**Rim** (54520057)	35—15	1	
8060	**Screw,** fixing Rim, (127511)	35—16	1	
8062	**Light Unit** (54520142)	35—17	1	
323	**Bulb** (323)	35—18	1	
8401	**LIGHT UNIT "RANGER" FOR FOG LAMP (54520077)**		1	TO CONVERT FOG LAMP TO LONG RANGE DRIVING LAMP
432011	**R.H. BRACKET FOR FOG OR LONG RANGE DRIVING LAMP**	35—19	1	

Part No.	Description	Plate No.	No. per Unit	Remarks
432012	**L.H. BRACKET FOR FOG OR LONG RANGE DRIVING LAMP**		1	
431963	**Grommet,** in Bracket for Lamp Cable		2	
UFS.125/6R	**Screw,** Set, securing Bracket to Bumper Spring Bar		4	
UFN.125/L	**Nut,** on Setscrews		4	
FG.104/X	**Washer,** Spring, under Nuts		4	
432009	**SWITCH, OPERATING FOG AND LONG RANGE DRIVING LAMP (34285)**	35—20	2	
431607/02	**Knob, 'F',** on Switch	35—21	1	
431607/03	**Knob, 'L',** on Switch		1	
4322511	**Bezel,** for Switch	35—22	2	

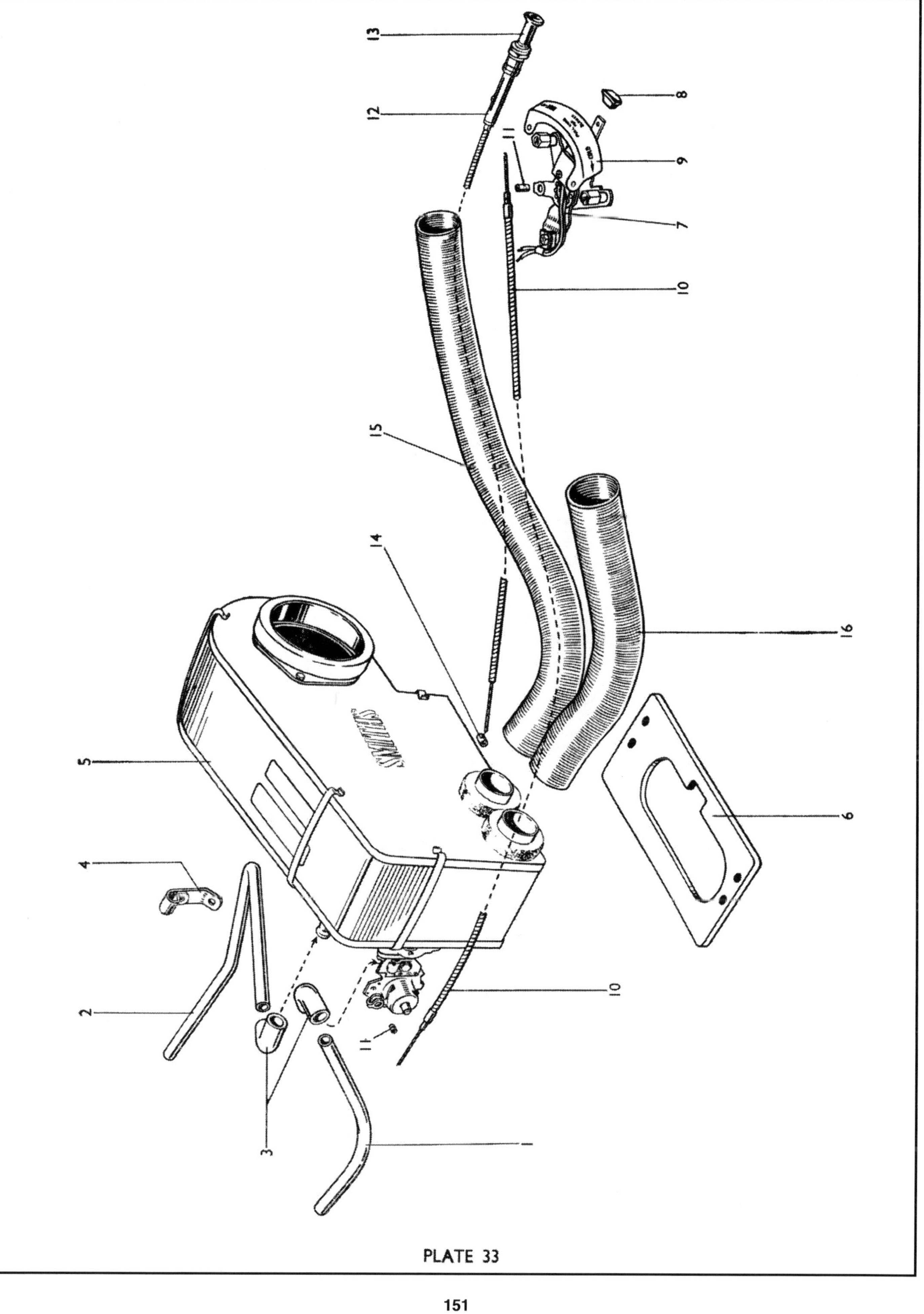

PLATE 33

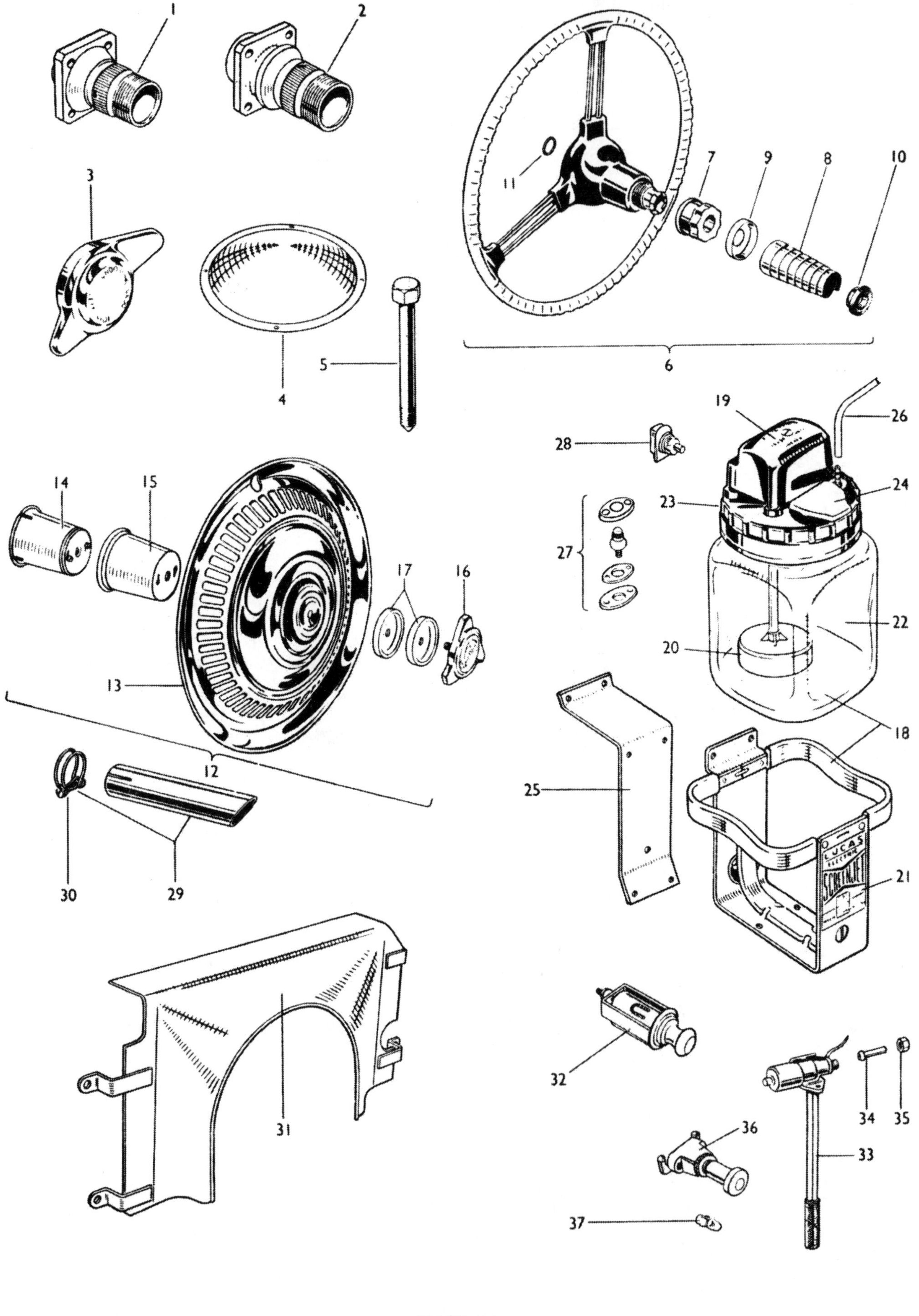
1
2
3
4
5
6
7
8
9
10
11
12
13
14
15
16
17
18
19
20
21
22
23
24
25
26
27
28
29
30
31
32
33
34
35
36
37
LUCAS

PLATE 34

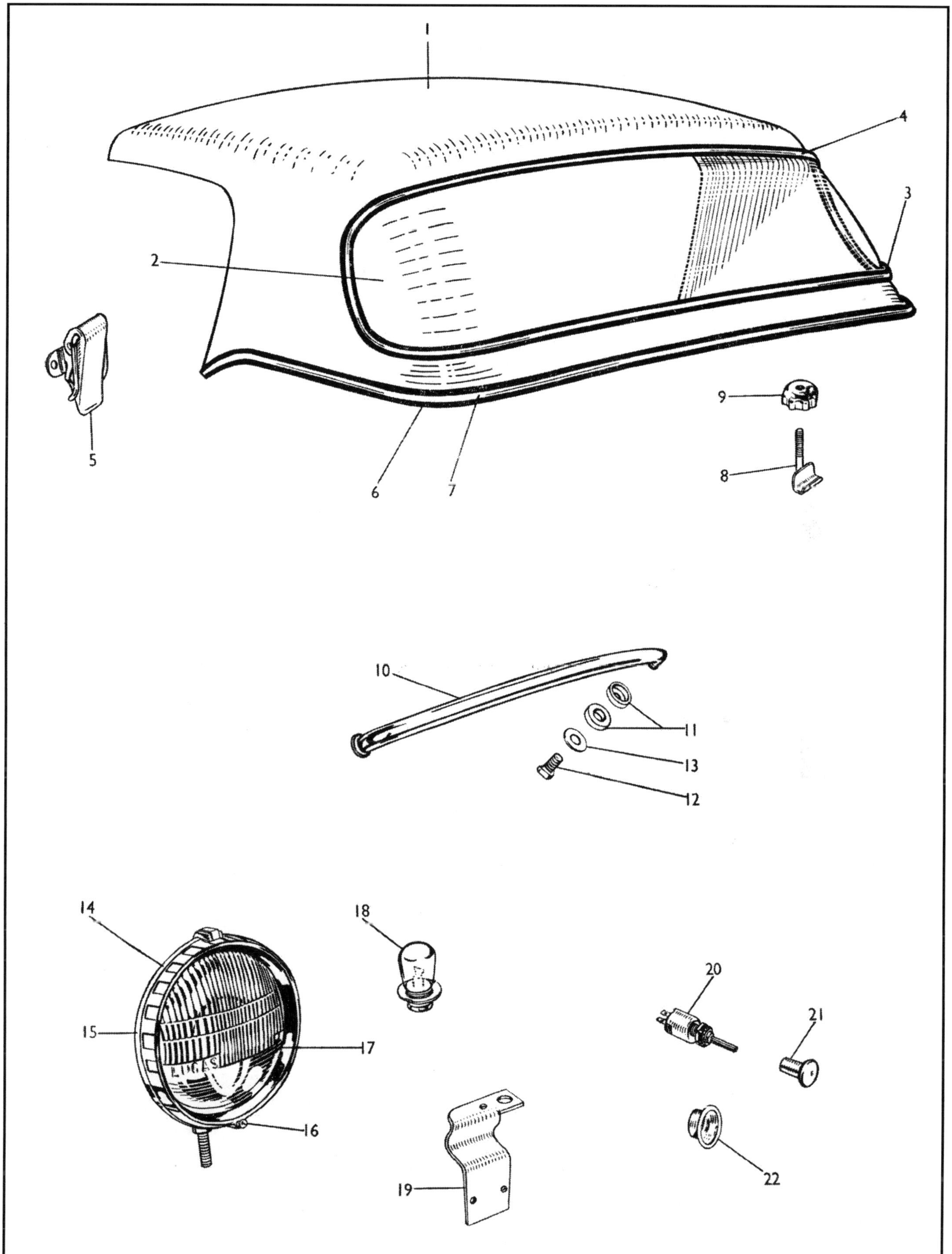

PLATE 35

Daimler 2.5 Litre V8 and 250 Saloon Service Manual

This Official Factory Service Manual covers the Daimler SP250 Sports car and 250 Saloon which is fitted with a 2.5 litre V8 engine. Detailed maintenance and repair procedures including, engine, cooling, fuel & exhaust, clutch, gearbox, brakes, suspension, steering & body and much more.
A total of 276 pages.

Publication No. E1002

ISBN 9781855208759

Daimler Dart SP250 & V-8 250 Saloon Road Test Portfolio

The Daimler Dart made its debut in 1959 and was then renamed the SP250. Some 2,645 were produced. Jaguar later acquired Daimler and in 1962 introduced the V-8 250 saloon sporting a Jaguar Mk. 2 body and the V-8 engine from the SP250, with production ending in 1969. All models are reported on. Included are road and used car tests plus advice on acquiring a good used example.
A total of 200 fully illustrated pages.

ISBN 9781855209145

Published by Brooklands Books Ltd., PO Box 904,
Amersham, Bucks HP6 9JA, England
brooklands-books.com

ISBN: 9781783181650 Ref: DSPPH Part No. D.5 12W5/3031

Printed in Dunstable, United Kingdom

91223684R00087